Quantitative Phase Microscopy and Tomography

Techniques using partially spatially coherent monochromatic light

Online at: https://doi.org/10.1088/978-0-7503-3987-2

IOP Series in Advances in Optics, Photonics and Optoelectronics

SERIES EDITOR

Professor Rajpal S Sirohi Consultant Scientist

About the Editor

Rajpal S Sirohi is currently working as a faculty member in the Department of Physics, Alabama A&M University, Huntsville, Alabama (USA). Prior to this, he was a consultant scientist at the Indian Institute of Science, Bangalore, and before that he held a chair and was professor in the Department of Physics, Tezpur University, Assam. During 2000–11, he was an academic administrator, being vice-chancellor to a couple of universities and the director of the Indian Institute of Technology, Delhi. He is the recipient of many international and national awards and the author of more than 400 papers. Dr Sirohi is involved with research concerning optical metrology, optical instrumentation, holography, and speckle phenomena.

About the series

Optics, photonics, and optoelectronics are enabling technologies in many branches of science, engineering, medicine, and agriculture. These technologies have reshaped our outlook and our ways of interacting with each other and have brought people closer. They help us to understand many phenomena better and provide a deeper insight into the functioning of nature. Furthermore, these technologies themselves are evolving at a rapid rate. Their applications encompass very large spatial scales from nanometers to astronomical and a very large temporal range from picoseconds to billions of years. This series on advances on optics, photonics, and optoelectronics aims to cover topics that are of interest to both academia and industry. Some of the topics to be covered by the books in this series include biophotonics and medical imaging, devices, electromagnetics, fiber optics, information storage, instrumentation, light sources, CCD and CMOS imagers, metamaterials, optical metrology, optical networks, photovoltaics, free-form optics and its evaluation, singular optics, cryptography, and sensors.

About IOP ebooks

The authors are encouraged to take advantage of the features made possible by electronic publication to enhance the reader experience through the use of color, animation, and video and by incorporating supplementary files in their work.

Do you have an idea for a book you'd like to explore?

For further information and details of submitting book proposals, see iopscience.org/books or contact Ashley Gasque on Ashley.gasque@iop.org.

Quantitative Phase Microscopy and Tomography

Techniques using partially spatially coherent monochromatic light

Dalip Singh Mehta
Department of Physics, Bio-photonics and Green Photonics Laboratory, Indian Institute of Technology Delhi, Hauz-Khas, New Delhi, India

Ankit Butola
The Department of Physics and Technology, UiT The Arctic University of Norway, Tromso, Norway

Veena Singh
Department of Physics, Bio-photonics and Green Photonics Laboratory, Indian Institute of Technology Delhi, Hauz-Khas, New Delhi, India

IOP Publishing, Bristol, UK

ISBN 978-0-7503-3987-2 (ebook)
ISBN 978-0-7503-3985-8 (print)
ISBN 978-0-7503-3988-9 (myPrint)
ISBN 978-0-7503-3986-5 (mobi)

DOI 10.1088/978-0-7503-3987-2

Version: 20221201

IOP ebooks

British Library Cataloguing-in-Publication Data: A catalogue record for this book is available from the British Library.

Published by IOP Publishing, wholly owned by The Institute of Physics, London

IOP Publishing, No.2 The Distillery, Glassfields, Avon Street, Bristol, BS2 0GR, UK

US Office: IOP Publishing, Inc., 190 North Independence Mall West, Suite 601, Philadelphia, PA 19106, USA

Contents

Preface

Quantitative phase microscopy (QPM) has become an indispensable technology for label-free high-contrast imaging and quantitative analysis of various biological cells and tissues, such as, sperm cells, micro-phases, liver sinusoidal cells, red blood cells, cancerous cells, plant cells etc, without using any exogenous contrast agents. Quantifying various bio-physical parameters of these cells and tissues is important for successful IVF technology, early-stage disease detection, hemoglobin concentration, sickle cell disease, dry mass determination, dynamic membrane fluctuations etc. Precise determination of these parameters depends on the accuracy in phase measurement with reduced spatial phase noise and high temporal phase stability. Most QPM techniques utilize highly coherent light like lasers due to their remarkable properties, such as high spatial and temporal coherence and brightness. But a laser's high spatio-temporal coherence leads to the occurrence of speckle noise and spurious fringes leading to poor spatial phase sensitivity and inhomogeneous illumination. All these factors significantly influence the phase measurement accuracy of QPM systems and lead to poor image quality.

During the last several years, QPM technology has been developed using partially spatially coherent monochromatic (PSCM) light and demonstrated an order of magnitude reduced spatial phase noise, high space bandwidth product, and free-from spurious fringes compared to using coherent lasers. It has also been demonstrated that by means of using common path interferometric systems a highly stable speckle-free QPM system has been realized so that one could achieve super stable fringe pattern leading to an order of magnitude higher temporally stable compared conventional off-axis systems. This system can be utilized for precise measurement of dynamic membrane fluctuation studies. The QPM systems with PSCM light sources lead to ~10-fold improvement in spatial-phase sensitivity and phase measurement accuracy compared to coherent laser light. By means of using PSCM with common path configuration one could also achieve ~10-fold improvement in temporal phase stability.

This book describes the most advanced QPM techniques and computational imaging techniques using partially spatially coherent monochromatic (PSCM) light rather than highly coherent lasers. It covers topics, such as, speckle-free QPM both off-axis and common path interferometric configurations, structured illumination phase microscopy (SIPM), chip-based nanoscopy, machine learning, deep learning, and artificial intelligence (AI) in phase microscopy. Optical coherence tomography (OCT) using longitudinal spatial coherence light as well as low temporal coherence light is also demonstrated. White light interference microscopy with color fringe analysis, multi-spectral and hyper-spectral phase microscopy is also demonstrated. This coherent-artifacts-free QPM leads to an order of magnitude improved spatial phase sensitivity, space-bandwidth product, and high temporal phase stability. The technique was utilized for sperm cells, red blood cells, macrophages, liver sinusoidal endothelial cells, and cancer cells. Recent advancements in speckle-free multi-spectral and hyperspectral QPM techniques are also discussed.

The text will be a useful to researchers, doctoral and post-doctoral students working in the area of biomedical optics, biophotonics, advanced microscopy, holography and optical metrology.

Acknowledgement

The authors are thankful to Shilpa Tayal, Priynaka Mann, Sunil Bhatt, Pamila Thapa, Virendra Kumar, Anuj Saxena and Sathi Das, Bio-photonics Group, Department of Physics, Indian Institute of Technology Delhi, India for their help from time to time during the preparation of this book.

The authors are also thankful to Dr Azeem Ahmad, Dr Vishesh Kumar Dubey and Professor Balpreet Singh Ahluwalia, from University of Tromso, Norway and their help and encouragement. The authors are also thankful to their family members for their encouragement.

One of the authors, Dalip Singh Mehta, is thankful to the Department of Physics, Indian Institute of Technology Delhi, India for encouragement to write books on this advanced topic.

Author biographies

D S Mehta

Dr Dalip Singh Mehta is currently a Professor at the Department of Physics, Indian Institute of Technology Delhi. Previously, he worked as Associate Professor and Assistant Professor (June 2002–December 2012) at the Instrument Design Development Centre, Indian Institute of Technology Delhi. Before Joining the Institute, he was JSPS Post-Doctoral Fellow (July 2000–June 2002) at the University of Electro-communications, Tokyo, Japan, Post-Doctoral Fellow National Dong Hwa University, Taiwan (November 1999–May 2000) Research Associate, NPL, New Delhi, STA Post-Doctoral Fellow (June 1997–May 1998), NIRE, Tsukuba, Japan, and UNESCO Research Fellow (January1996–September 1996), Tokyo Institute of Technology Tokyo, Japan. His research areas are: bio-photonics and biomedical optical imaging: optical coherence tomography, label-free quantitative phase microscopy and optical tweezers; green photonics: laser based solid state lighting/VLC/Li-Fi and sunlight harvesting for daylight saving; optical metrology, optical 3D-surface profilometry, optical interferometry and optical coherence. He has contributed 180 research papers in international refereed journals, and more than 200 in international and national conferences. He has delivered more than 75 invited talks/lectures in various international and national conferences and universities. He has supervised 21 PhD students, is currently supervising 16 PhD students and has filed 16 patents.

Ankit Butola

Ankit Butola received his PhD in optics and photonics from Indian Institute of Technology Delhi, India in 2021. He is a postdoctoral fellow at the Department of Physics and Technology, UiT, The Arctic University of Norway. His research interests are at the crossroads of optics, physics, computer science and biology. His research interests are microscopy, bio-imaging, machine learning and optical nanoscopy.

Veena Singh

Veena Singh received her BTech degree in the field of electronics and communications from University Institute of Engineering and Technology, CSJM University, Kanpur in 2010. She completed her MTech. in laser technology from IIT Kanpur in 2012. Thereafter, she joined the PhD program at Department of Physics, IIT Delhi in July 2014 and worked in the areas of bio-photonics and applied optics. The work carried out by her has been published in peer-reviewed and high reputed journals. She has authored/co-authored in 10 journal articles and 19 conference proceedings. The title of her PhD thesis is 'Development of highly-stable, wide-field and high-resolution quantitative phase microscopic system for biological applications'. She has hands-off experience in the fabrication and designing of optical systems.

Chapter 1

Introduction

In this chapter we first briefly describe bright field microscopy, its magnification, lateral and axial resolution followed by phase contrast microscopy and quantitative phase microscopy (QPM). Since most of the QPM techniques are implemented using highly coherent laser light, therefore, the coherence (both spatial and temporal) properties of laser light are described including the calculations of coherence time and coherence length of various sources of light. The high spatio-temporal coherence of laser light generates speckle and spurious fringes, therefore the origin of speckles and their effects in imaging, QPM and space-bandwidth product of QPM are illustrated. Finally, various methods of generating partially spatially coherent light for QPM are listed.

1.1 Bright field microscopy

Optical bright field microscopy has been an important technique to visualize the microstructures of biological specimens [1–3]. The details of fine structure in micro-organisms have been visualized with the help of high-resolution optical microscopes. Since most of the bright field microscopes are implemented by means of using electromagnetic radiation in the visible range, i.e., white light (450–780 nm), they are harmless to living biological cells and micro-organisms. Further, it is a non-contact and non-invasive technique, and therefore most friendly equipment for living biological specimens. If the biological specimens are highly scattering or absorbing objects then bright field optical microscopes can produce good contrast images of the specimens, therefore, fine details of the specimen such as, cell walls, nucleus, and cell boundaries can be visualized easily. Optical bright field microscopes both in transmission and reflection mode have been developed and widely used. For biological applications mostly transmission mode is preferred, while for reflecting objects and surfaces, particularly in material sciences, the reflection mode microscope is used. Figure 1.1 depicts the schematic ray diagram of the bright-field compound microscope system in transmission mode. Collimated light is focussed by

doi:10.1088/978-0-7503-3987-2ch1

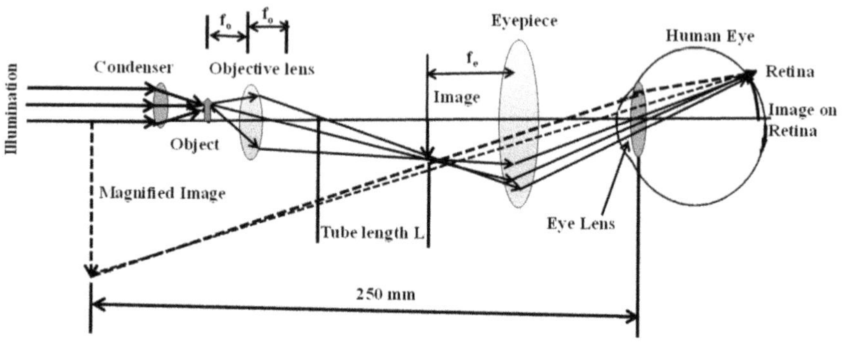

Figure 1.1. Illustrating the schematic ray diagram of bright-field microscope in transmission mode.

the condenser lens onto the object and the object scatters light in all directions. The scattered light in the forward direction is collected by the microscope objective lens. If the object is placed at the focus of the lens then a magnified inverted image is formed inside the tube, as shown in figure 1.1.

The microscope objective (MO) forms a real, inverted, and magnified image of the micron size object. This image is formed before the eyepiece at the primary focal plane. It further magnifies the intermediate image which is formed onto the retina. Thus the magnifying power of the entire microscopic system is the product of the lateral linear magnification of the objective and the magnification of the eyepiece lens. Total linear magnification of the above compound microscope is $M = M_{\text{Obj}}. \times M_{\text{Eyepiece}}$, where M_{Obj} and M_{Eypiece} are the magnifications of objective lens and eyepiece, respectively. Typically, the maximum value of M is $100\times \cdot 10\times = 1000\times$. The magnification also depends on the tube length of the microscope. i.e., the larger the tube length the higher the magnification [1–3].

1.1.1 Lateral (transverse) resolution

The resolution of the microscope can be defined as the smallest object/distance between two points on to the object that can be distinguished as two separate entities. The lateral (transverse) resolution of the microscope is limited by the diffraction phenomenon due to the wave nature of light. The diffraction limited resolution of the entire compound microscope is given by the following formula [1–3];

$$R = \frac{1.22n\lambda}{NA_{\text{Objective}} + NA_{\text{Condenser}}} \tag{1.1}$$

where 'n' is the refractive index of the sample and λ is the wavelength of the illuminating light. In many microscopic systems particularly in reflection mode microscope in general $NA_{\text{Objective}} = NA_{\text{Condenser}}$, therefore, the above equation is reduced to;

$$R = \frac{0.61n\lambda}{NA_{\text{Objective}}} \tag{1.2}$$

As can be seen from equation (1.2), the higher the value of NA and the smaller the value of λ, the higher is the resolution of the microscope. The maximum value of NA for a dry objective lens is 1.0 but for oil immersed objective lenses the NA can be improved up to 1.40.

1.1.2 Axial (longitudinal) resolution

The axial resolution of the compound microscope can be given by the following expression [1–3];

$$Z_{\min} = \frac{2n\lambda}{(NA_{\text{Objective}})^2} \tag{1.3}$$

The resolving power of the microscope is defined as the ability to resolve two close object points and is given by the inverse of the equation (1.2).

Figure 1.2 shows the schematic diagram depicting the diffraction limited point spread function of the light after passing through a microscope objective lens. It can be seen that the resolution of a wide-field microscope is poor, i.e., in lateral direction it is ~200 nm and in axial direction it is ~500 nm, therefore, object points closer to these values cannot be resolved. The number of red colour circles in the lateral and axial direction shows that if the object contains nanoparticles of size 50 nm each then they cannot be resolved using this bright field microscope. Further, the images are blurred, i.e., fine details of the object are not discernible, contrast is poor (low signal-to-noise ratio) due to high background noise, no optical sectioning, 3D-imaging, and no quantitative information can be obtained. These are the main limitations of the wide-field microscope.

However, many living cells or tissues are transparent under visible light illumination, i.e., they do not absorb (or scatter) light significantly [4–6]. Thus, a magnified image of cells or tissues will generally exhibit low contrast. To solve this problem, fluorescence microscopy was developed in which an exogenous (extrinsic) contrast agent, such as organic dyes, are used, which are excited by the excitation light, generally green, blue, and UV light source. The exogenous contrast agents

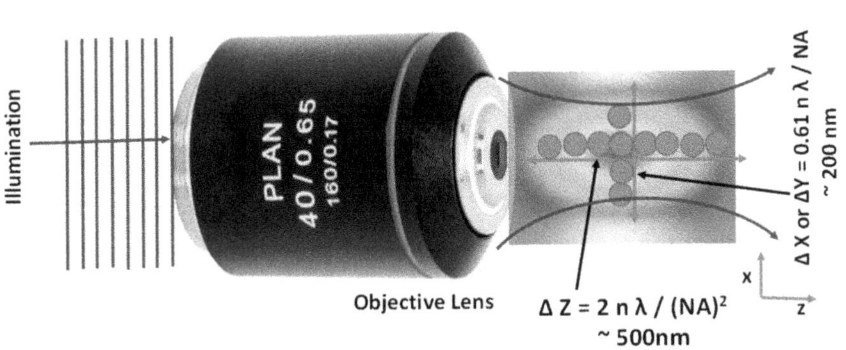

Figure 1.2. Schematic diagram illustrating the diffraction limited point spread function.

bind at sundry sites in the cell and absorb excitation light and the dye molecules are raised to high energy state and then spontaneously emit low energy (long wavelength) photons called fluorescence which render the structures visible with contrast. Fluorescence microscopy is the most used exogenous contrast enhancing technique in cell biology, which not only provides high contrast in cell imaging but also provides specificity by labelling specific structures with certain fluorophores and has been a very powerful technique. Although the contrast has improved in fluorescence microscopy, it nevertheless still suffers from diffraction limited resolution and blurring. Further, using an exogenous contrast agent may lead to toxicity and photo-bleaching. In many delicate samples it is better to avoid the use of exogenous contrast agents and use only bright-field microscopy to visualize the samples. Phase contrast microscopy was invented [6–10] to visualize the fairly transparent objects which do not absorb or scatter light strongly.

1.2 Phase contrast microscopy

Phase contrast microscopy (PCM) is label-free high contrast imaging of transparent biological cells or tissues using visible light, which was first invented by F Zernike in 1930 [6]. In 1948, Danish Gabor [7, 8] invented a new microscopic principle without any lenses, which is known as holographic microscope. In both the techniques without using any exogenous contrast agents, the amplitude and phase of object images were reconstructed using interferometric principle. Many biological cells and tissues are transparent to visible light, i.e., they do not absorb or scatter light strongly and hence the magnitude of light transmitted through the cells does not change significantly, however, the phase of the incident light changes significantly due to the change in optical path length (OPL = refractive index $n(x, y) \times$ geometrical thickness $t(x, y)$) across the cell, i.e., cells are *phase objects* [6–8]. Due to this change in OPL the light travels with different speeds in different regions of the cell/tissues, i.e., that light experiences different OPL (*phase*) delays across the cell. In PCM this change in OPL or phase shifts in the light transmitted through the specimen are converted into significant changes in amplitude or contrast value across the image, thus visualizing the fine details of the objects, such as cell walls, boundaries, nucleus etc. This is possible by means of constructive and destructive interference of light across the sample, therefore, no exogenous contrast agent or no staining is required. Figure 1.3(a) shows the schematic of contrast improvement by means of using an exogenous contrast agent, in general organic dye infiltrated. In this excitation light vibrations interact with a biological specimen tagged with fluorescent dye. The dye molecules absorb excitation blue light and emit green fluorescent light. This way one can improve the contrast of the image. Figure 1.3(b) shows the contrast improvement in PCM. In this the amplitude of the light vibrations, illuminating the samples least absorbed or scattered by significant phase change takes place. Therefore, in some locations the light vibrations are out of phase and in some locations they are in constant phase difference or in phase, thus leading to constructive or destructive interference. The spatial variation of this interference pattern leads to significant change in contrast also without using any exogenous

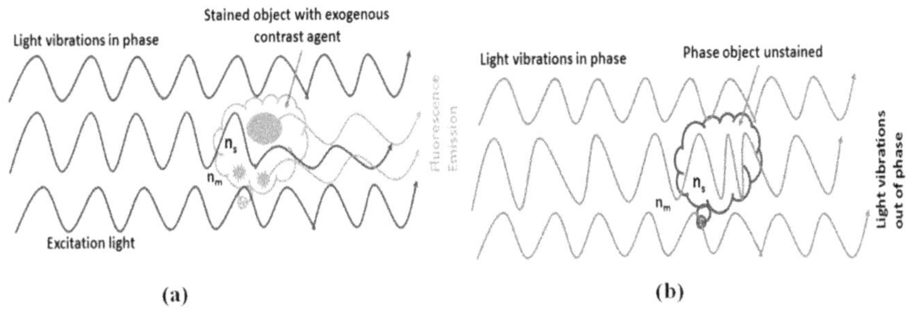

Figure 1.3. (a) Schematic diagram illustrating excitation light vibrations in phase interacting with a stained object with exogenous contrast agent and emitted fluorescence. (b) Light vibrations in phase interacting with phase object unstained.

contrast agent. In PCM the phase transmittance function of the light vibrations after passing through the sample is given as [6–8]:

$$T(x, y) = \exp[j\varnothing(x, y)] \cong 1 + j\varnothing(x, y) \tag{1.4}$$

In PCM an annular beam of light is first generated and then focussed via a condenser lens onto the sample and it scatters the light [6, 7]. The objective lens collects both the scattered and unscattered beams and passes through a phase plate to introduce the desired (π/2) phase between the two. Finally interference takes place onto the detector plane. Details about the schematic diagram and experimental set-up can be found ealsewhere [6–8]. Later on, a differential interference contrast (DIC) microscope was developed [9, 10]. In this scheme the linearly polarised light passes through a Wollaston prism first which generates two beams, one S-polarized and another P-polarised. One of the beams passes through a biological sample and another beam without sample. Another Wollaston prism was used to combine both the beams. When the state of polarization of the beam passing through samples changes, the interference takes place. DIC microscope has also been using rigourously for contrast improvement and determining the polarization state of the sample.

1.3 Quantitative phase microscopy

Both PC and DIC microscopy provide only improved contrast and qualitative information about the sample. Quantitative phase microscopy (QPM) was developed [11–15], which is a label-free high-contrast modality for transparent or semi-transparent biological specimens. Quantifying the optical path lengths, i.e., phase shifts associated with the cells and tissues gives information about morphology and dynamics at the nanometer scale. Further, determination of refractive index of the cells, tissues, 3D-shape, structure, and height map can be obtained using QPM. Study of highly dynamic phenomena such as cell growth, cell membrane fluctuations and cell motility are possible using QPM. Dynamic membrane fluctuations of sickle red blood cells (RBCs) are significantly lower than healthy RBCs, leading to reduced cell deformability. So far, different

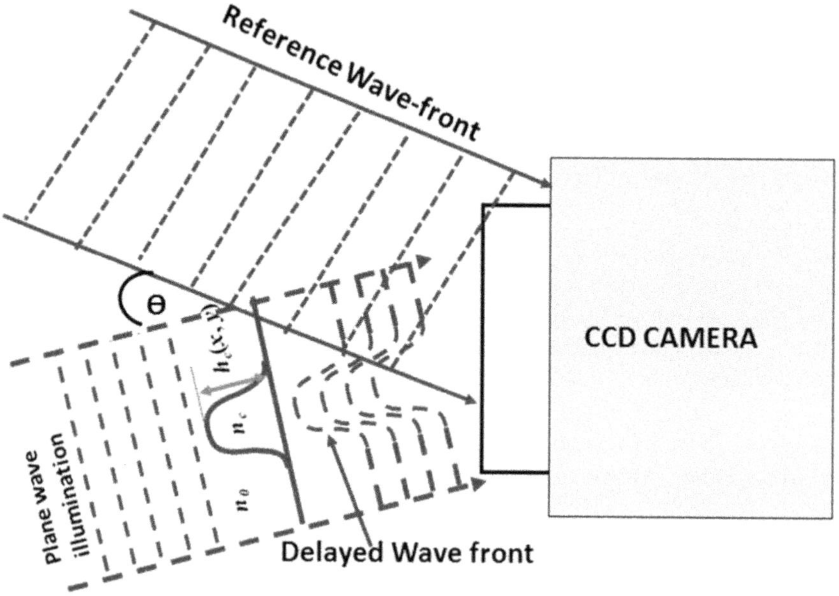

Figure 1.4. Illustrating the concept of QPM with interference of object and reference waves.

techniques such as digital holographic microscopy [13, 14], Fourier phase microscopy [16], Hilbert phase microscopy [17, 18], diffraction phase microscopy [19–21], spectroscopic diffraction phase microscopy [22, 23], tomographic phase microscopy [24], phase shifting white light interference microscopy [25–28] and spectrally resolved laser interference microscopy [29] have been developed. Each of these techniques has been highly popular and has its own advantages and disadvantages. In all the above QPM techniques a coherent light source like laser is used and the single beam is divided into two beams, one serves as reference beam and another beam passes through the biological specimen and is called the object beam (figure 1.4). Both the beams are superposed and made to produce interference. The 2D-interferogram is then recorded by solid-state area detectors like charge coupled devices (CCD)/complementary metal oxide semiconductor (CMOS) and stored in the computer for further processing. A suitable computer algorithm is used to reconstruct the phase map of the object.

The intensity of the 2D-interferogram can be expressed as [11, 12];

$$I(x, y) = I_R(x, y) + I_O(x, y) + 2\sqrt{I_R(x, y)I_O(x, y)}\cos[\Delta\varphi(x, y)] \tag{1.5}$$

where

$$\Delta\varphi\,(x, y) = \underbrace{2\pi/\lambda \times 2h_c(x, y) \times (n_c(x, y) - n_0)}_{\text{OPD(optical path delay) profile} \sim \text{Thickness} \times \text{Integral refractive index}} \tag{1.6}$$

where $I_R(x, y)$ and $I_O(x, y)$ are the intensities of light passing through object and reference beams, respectively, $\Delta\phi(x, y)$ is the phase difference between two light beams, $h_c(x, y)$ is the height variation of the cell or tissue, $n_c(x, y)$ refractive index distribution of the cell and n_0 is the refractive index of surrounding medium. By means of recording the five phase shifted interferograms, the phase map $\Delta\phi(x, y)$ can be reconstructed with high accuracy and $n_c(x, y)$ refractive index distribution of the cell can be determined. Further, the Barer's expression for refractive index of RBC is related to various biophysical parameters of the cell via the following expression [7, 29, 30]

$$\Delta n_c(x, y) = \alpha(\lambda)C(x, y) + n_X(x, y, \lambda) \tag{1.7}$$

where $\Delta n(x, y)$ is refractive index distribution of RBC, $\alpha(\lambda)$ is the refractive index increment, $C(x, y)$ the cytoplasmic Hb concentration in the RBC and $n_X(x, y, \lambda)$ is the average refractive index of other molecules besides Hb in RBC. QPM has the capability to quantify various biophysical parameters of the cells, such as sample thickness, refractive index (n), cell dry mass, haemoglobin (Hb) concentration etc. These parameters are determined by means of measuring the distorted phase map of the light field interrogating with the sample using interferometric or holographic techniques along with suitable reconstruction algorithms. Considering RBCs cytoplasm to be constant, the height of an RBC can be given by [31, 32]

$$h_c(x, y) = \frac{\lambda}{2\pi(n_{RBC}(x, y) - n_r)} \cdot \varphi(x, y) \tag{1.8}$$

$$\varphi(x, y) = \frac{2\pi}{\lambda}(n_{RBC}(x, y) - n_r)h_c(x, y) \tag{1.9}$$

Equations (1.8) and (1.9) give the direct relationship between refractive index and Hb concentration. Thus, knowing the refractive index of RBC plays an important role in determining the various properties of the sample as whole blood which mostly comprises erythrocytes as well as plasma; erythrocytes being 40%–45% and plasma being 55%–60%. The refractive index of whole blood depends on the refractive index of plasma as well as on the refractive index of RBCs. The refractive index of RBCs relates to the internal property of the sample and can act as the indicator of the health of a person. The progression of any disease can be directly related to changes in the morphology and refractive index of the cell, thereby making the measurement of refractive index as an important parameter to focus on. Presently, the QPM has become one of the most indispensable methods in the field of modern healthcare and disease diagnosis via quantitative measurement of the above. Most of the QPM techniques are generally employed with highly coherent light sources like lasers due to their remarkable properties, such as high spatial and temporal coherence, high brightness, directionality and high colour purity/monochromaticity. Due to the long temporal coherence length of lasers, it is very easy to realize an interference microscope. The following section describes the coherence properties of laser light.

1.4 Coherence properties of laser light

1.4.1 Temporal coherence properties

There are two fundamental processes of light emission from sources, spontaneous and stimulated emission. Lasers emit light via stimulated emission process, whereas other sources emit light via spontaneous emission [33–36]. Due to stimulated emission in lasers the light waves emerging from the cavity are in phase (coherent), monochromatic, same polarization and same direction. Therefore, the lasers are highly spatially and temporally coherence light sources. On the other hand the non-laser sources are incoherent due to spontaneous emission process. The lasers generate temporally highly coherent (monochromatic) light and the coherence time τ_c of laser light can be defined as the time duration within which the light vibrations are perfectly sinusoidal for finite time duration [33–36]. Figure 1.5(a) shows the coherence time τ_c of a highly monochromatic light, i.e., single frequency light source, which is an ideal source practically not realizable. Due to single frequency the light vibrations are perfectly sinusoidal for infinite time [36]. However, light generated by lasers has finite spectral bandwidth, although in some cases extremely narrow bandwidth, hence can be termed as quasi-monochromatic (QM), $\Delta\nu \ll \nu_0$ [33–35]. Such type of light sources generate light which is perfectly sinusoidal for finite time duration and coherence time is also finite as can be seen from figure 1.5(b). The coherence time can be computed using the relation $\tau_c = 1/\Delta\nu$ [33–36]. In general, to realize experiments the coherence is computed in space domain and the coherence length is computed by $l_c = c \times \tau_c$. The coherence length l_c of the light source can be defined as the path length within which the light vibrations are perfectly sinusoidal (figure 1.5(c)) or two light vibrations arriving at same spatial position P at two different paths l_1 and l_2 are in phase or have constant phase difference then they interfere constructively. The coherence length l_c can be further expanded as follows [33–36];

$$l_c = c \times \tau_c = c/\,\Delta\nu = \lambda^2/\,\Delta\lambda;\ \ (\text{using } \nu = c/\lambda;\ \Delta\nu = c \times \Delta\lambda/\lambda^2) \qquad (1.10)$$

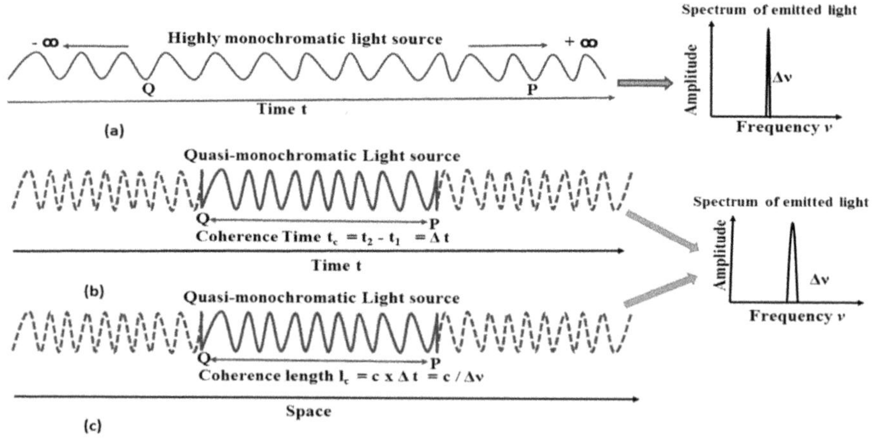

Figure 1.5. Illustrating the temporal coherence of laser light; (a) highly monochromatic light source, (b) quasi-monochromatic light source in time-domain; (b) same as (a) in space domain.

The temporal coherence of the light source is defined as the fixed or constant phase relationship, i.e., correlation between the light vibrations at two different instants of time but at fixed spatial position. According to Wiener–Khintchin theorem, auto-correlation or temporal coherence function $\Gamma(\Delta t) = \langle E(t)E^*(t - \Delta t)\rangle$ and source power spectral density forms Fourier transform pairs and is given by the following expression [34, 37].

$$\Gamma(\Delta t) = \int_{-\infty}^{\infty} S(\nu)\exp(i2\pi\nu\Delta t)d\nu \tag{1.11}$$

where, $\Gamma(\Delta t)$ is the temporal coherence (TC) function, $S(\nu)$ is the source spectral distribution function, and Δt is the temporal delay between optical fields $E(t)$ and $E^*(t - \Delta t)$. According to the Wiener–Khinchin theorem, the TC function and the source spectral distribution form the Fourier transform pairs (see equation (1.11)) [34, 37]. In other words, the larger the bandwidth of source temporal frequency spectrum, the smaller will be the coherence length of light source or vice versa. Therefore, broad-band frequency spectrum light sources, such as, tungsten halogen lamps, super-luminescent diodes (SLDs), and broad-band Ti:sapphire lasers are used for obtaining high-axial resolution sectioning in optical coherence tomography (OCT) imaging [38, 39]. In general, the spectral profile of the light sources is Gaussian, therefore, equation (1.11) can be re-written as:

$$l_c = \frac{2\ln 2}{\pi}\frac{\lambda^2}{\Delta\lambda} \tag{1.12}$$

The temporal coherence length of the light source is inversely proportional to the full width at half maximum (FWHM) ($\Delta\lambda$) of the source spectrum. Therefore, the larger the $\Delta\lambda$, the smaller is the coherence length of the light source. In the case of polychromatic (broad-band) light sources the coherence length is extremely small and bandwidth is large. Figure 1.6 shows the coherence length of a polychromatic

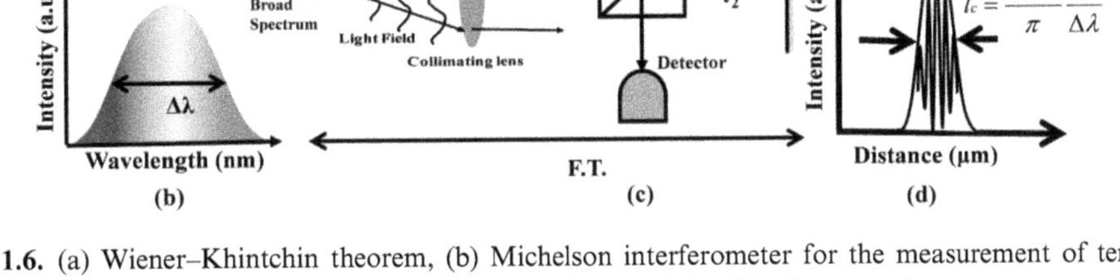

Figure 1.6. (a) Wiener–Khintchin theorem, (b) Michelson interferometer for the measurement of temporal coherence of light source, (c) and (d) Fourier transform relationship between the source spectrum and coherence function of light source.

(broad-band) light source. The typical examples of polychromatic (broad-band) light sources are white light LEDs, SLD and a super-continuum laser source (figure 1.7).

Table 1.1 shows the peak wavelength, spectral band, coherence time, coherence length of lasers and non-laser light sources which are commonly used in interferometry, holography, optical metrology, fluorescence microscopy and in bright-field microscopy. Single-mode lasers (S. Nos. 1 and 2, in table 1.1) have longer coherence time and long coherence length compared to other light sources [34, 35] and hence they are frequently used for LIGO, interferometry, holography and QPM. On the other hand, short coherence length light sources (S. Nos. 4–8 in table 1.1) are used for low coherence interferometry, profilometry, OCT.

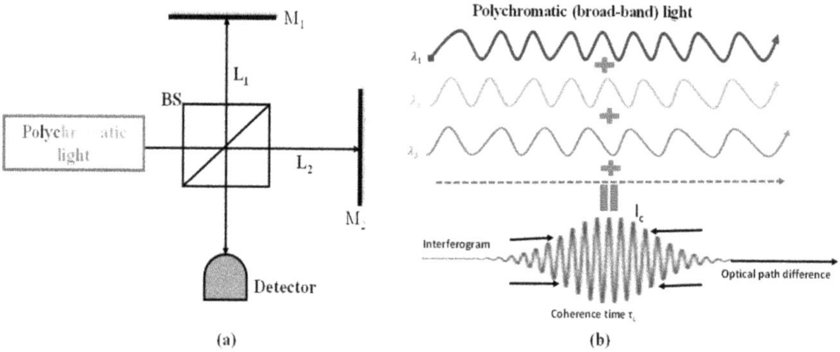

Figure 1.7. (a) Michelson interferometer and (b) illustrating the coherence time and coherence length of a polychromatic (broad-band) light source.

Table 1.1. Calculations of coherence time, and coherence length of various light sources [33, 35].

S. No.	Light sources	Peak wavelength λ_0	Spectral bandwidth ($\Delta\nu$)	Coherence time ($\tau_c = 1/\Delta\nu$)	Coherence length $l_c = c \times \tau_c \frac{2 \ln 2}{\pi} \frac{\lambda^2}{\Delta\lambda}$
1	Single-mode, frequency stabilized He–Ne laser	633 nm	1.0 KHz	1.0 ms	300 km
2	Single-mode DPSS laser	532 nm	1.5 MHz	0.667 μs	200 m
3	Diode laser	650 nm	3.5 GHz	0.286 ns	8.57 mm
4	Single colour LED	650 nm	1.50×10^{13} Hz	66.7 fs	20.0 μm
5	Super-luminescent diode (SLD)	840 nm	5.0×10^{13} Hz	20 fs	6.0 μm
6	White light source	550 nm	3.0×10^{14} Hz	3.33 fs	~1.00 μm
7	Filtered sunlight	550 nm	3.75×10^{14} Hz	2.67 fs	800 nm
8	Super-continuum laser	800 nm	1.50×10^{15} Hz	0.667 fs	200 nm

1.4.2 Spatial coherence properties of laser light

The spatial coherence of laser light is very high due to stimulated emission and size of the beam is small. Spatial coherence can be defined as, if the light vibrations at any two spatial points P and Q (figure 1.8(a)) arriving at the same time, *t*, are in phase or having constant phase difference between them, then the light vibrations are called spatially coherent. Figure 1.8(a) shows a schematic diagram depicting high spatial coherence of laser light. Due to stimulated emission, high monochromaticity, positive feed-back and small size (~1.5 mm–2.0 mm) of the laser beam, all the light vibrations are in phase across the beam area, as shown in figure 1.8(a). This guarantees high directionally of the laser beam, which is one of the important properties of laser light compared to spontaneously emitted coherent light. Further, due to high directionality, the brightness (no. of photons per unit time and per unit area) of the laser light across this beam is very high. Due to high directionality and high brightness, the laser beam can be tightly focussed to a small spot size, which is important for single-beam optical tweezers. Due to high directionality, it is also very easy to align an interferometer system such as, Michelson or Mach–Zehnder interferometer and off-axis holographic system. On the other hand, figure 1.8(b) shows the incoherent light via spontaneous emission. The emitted light is both temporally as well as spatially incoherent. The phase of light vibrations is completely mismatched and omnidirectional and hence not at all directional. Figure 1.8(c)

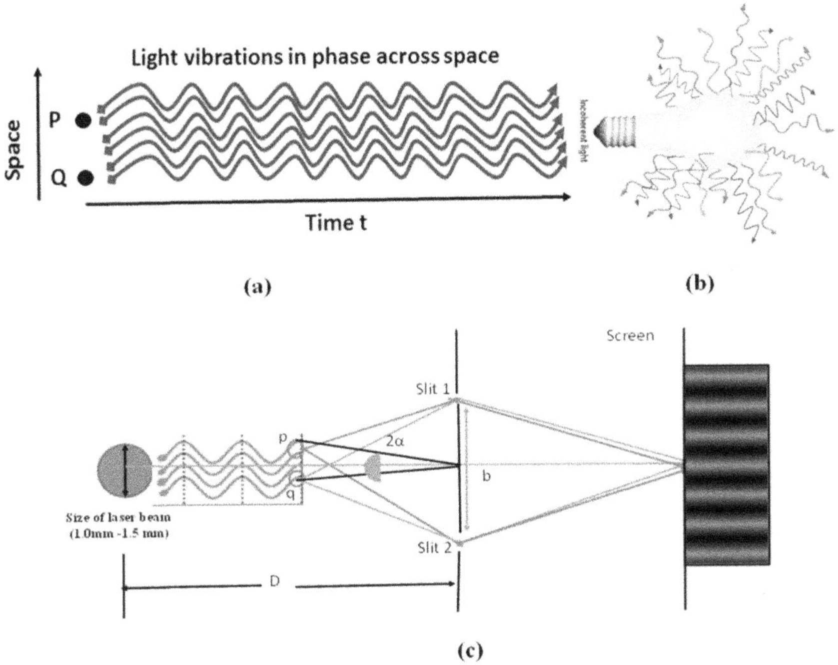

Figure 1.8. (a) Depicting spatially highly coherent laser light due to stimulated emission; (b) incoherent polychromatic light due to spontaneous emission; and (c) Young's double slit experiment for measuring spatial coherence of laser light.

shows the experimental set-up for the determination of spatial coherence and coherence area of laser light sources. In general, the size of the laser beam at the exit of lasers is around 1.0–1.5 mm, and the light is emitted via a stimulated emission process, therefore, the light vibrations across the area of the beam are in phase and hence the degree of spatial coherence is high. This can be proved by means of conducting the Young's double slit experiment and if the two slits p and q are within the diameter of the beam then a high visibility interferogram will be obtained as can be seen from the interferogram figure 1.8(c). [33–35].

Due to laser's high spatial coherence, high directionality, and large temporal coherence length, a QPM system based on holographic or interferometric optical configurations can be realized easily even though the optical path differences (OPDs) between the two beams are large enough and high contrast interference fringes can be obtained quickly. However, the major problems encountered in QPM techniques while using laser as a light source are the occurrence of spurious fringes due to multiple reflections from various optical components, strong speckle noise due to roughness of the sample and inhomogeneous illumination [31, 40–59].

All these factors affect the accuracy in phase/height measurement and resolution and lead to poor image quality in the laser based QPM. More recently there has been significant progress in the development of speckle-free QPM using partially spatially coherent light (PTLS) with reduced spatial phase noise and speckle-free imaging as compared to using a coherent laser directly. However, the key problems encountered with laser based QPM are the occurrence of parasitic/spurious fringes due to multiple reflections from different surfaces or interfaces of the optical components, strong speckle noise and inhomogeneous illumination. This is due to high spatial and temporal coherence properties of laser light source. All these factors significantly influence the phase measurement accuracy of QPM system and also lead to poor image quality [31, 40–59].

1.5 Origin of speckles in imaging and microscopy

Speckles are usually generated due to the interrogation of highly coherent electromagnetic radiation with rough surfaces. Visible light lasers are the best example to understand the origin of the speckle phenomena [46–48]. When a highly coherent light from a laser is back-reflected, scattered or transmitted through some rough surface, the light waves mutually interfere, resulting in constructive and destructive manner leading to the fine granular pattern known as laser speckles. The origin of speckles is a consequence of highly coherent intrinsic properties of the laser sources. Most of the objects in the world have the surface roughness of the order of the optical wavelengths. When a highly coherent light reflects or transmits through such surfaces, it scatters in the space which causes the optical path difference between the light rays and hence a corresponding phase difference is introduced between the various rays of the scattered light. The reason for this is that the mutual interference takes place to originate a speckle pattern on a screen or retina of the human eye. In this consequence, the observer realizes randomly distributed dark and bright granular patches of the light. The scattered light field with speckled noise has a random phase component (RPC). The resultant field **A** at the point of observation is calculated as a sum of RPC [60–63].

$$\mathbf{A} = Ae^{j\theta} = \frac{1}{\sqrt{N}}\sum_{n=1}^{N} a_n e^{j\varnothing_n} \tag{1.13}$$

where A is the magnitude of complex field. N denotes the number of random phasor components, a_n and $\varnothing_n$ imply the amplitude and the phase of nth complex phasor components. The light field intensity I is calculated as [60–63]:

$$I = \begin{cases} |A_x|^2 + |A_y|^2 & \text{for an unpolarized wave} \\ |A|^2 & \text{for a polarized wave} \end{cases} \tag{1.14}$$

The variance of the resultant phasor is given as [60, 61]:

$$\sigma^2 = \frac{1}{N}\sum_{n=1}^{N}\frac{a_n^2}{2} \tag{1.15}$$

Intensity fluctuations of light in a speckle pattern are measured in terms of speckle contrast C as a ratio of standard deviation to the mean of light intensity in a speckle pattern as [60–63]:

$$C = \frac{\sigma_I}{\langle I \rangle} = \frac{\sqrt{\langle I^2 \rangle - \langle I \rangle^2}}{\langle I \rangle} \tag{1.16}$$

The speckle contrast is essentially affected by the various parameters, i.e., spatial diversity, temporal diversity, polarization of light, angle of inclination, angle of detection and wavelength of the laser source. Various strategies of laser speckle reduction have been developed since the invention of lasers [62, 63]. An expanded laser beam passes through a diffuser scatter spatially and if it is incident onto another stationary diffuser, then a bucket integration of light photons takes place on the detector plane and results in a significant reduction in the speckle contrast. This strategy of speckle noise reduction is known as spatial diversity. Similarly, temporal diversity is another method of speckle reduction. Temporal diversity can be created by means of various methods, i.e., by passing the laser light through a rotating ground-glass plate, temporal variation of scattering media or by a vibrating diffuser screen etc. When a coherent beam of laser light transmits through these temporally varying mediums, the light photons are randomly scattered and arrive on the screen with a different time-lag. It happens much faster than the detector's response time. The multiple speckles add on intensity basis at the detector plane and lose their phase information completely [63–65]. Hence, a significant reduction in speckle noise can be observed. It is a most effective technique of laser speckle reduction, but this technique requires mechanical movements. When a linearly polarized light is incident on some objects of birefringent material or surface, in which the state of polarization of light is different at different positions. If a highly coherent light is allowed to pass through such a randomly polarized surface, then after reflection or transmission the light gets randomly polarized in two orthogonal directions. In both directions light intensities are equal and do not depend to each other. Such kind of polarization diversity can also be used for reduction of speckle contrast. The amount of reduction of speckle contrast through polarization diversity depends on the degree of polarization p of the light in such a way that speckle contrast [59–63],

$$C = \sqrt{\frac{1 + p2}{2}} \tag{1.17}$$

So, in this case the speckle contrast reduction is limited by a factor $\frac{1}{\sqrt{2}}$. If two laser sources are placed at an angle such that the light incident from these sources makes the angle of inclination $+\alpha$ and $-\alpha$ with the diffuser and if the angle of separation between the sources is large enough, i.e., $>2\alpha_{\min,}$ the speckle contrast will be reduced $>\frac{1}{\sqrt{2}}$ due to the angular diversity. Phenomena of interference of light depend on the source wavelength. As laser speckles are generated by the mutual interference of highly coherent light, speckle patterns must depend on wavelength of the laser light. If an average relative phase shift created by diffuser object is ~2π or more, then speckle patterns formed by two lasers of different wavelengths become uncorrelated. This method of laser speckle reduction is called wavelength diversity.

1.6 Effects of speckle on imaging and microscopy

Precise determination of various biophysical parameters using QPM depends on the accuracy in phase measurement with reduced spatial phase noise and high temporal phase stability. Most of the QPM techniques are generally employed with highly coherent light sources like lasers due to their remarkable properties, such as high spatial and temporal coherence, high brightness, directionality and high colour purity/monochromaticity. Due to the long temporal coherence length of lasers, it is very easy to realize an interference microscope. However, the key problems encountered with laser based QPM are the occurrence of parasitic/spurious fringes due to multiple reflections from different surfaces of the optical components, strong speckle noise and inhomogeneous illumination. This is due to high spatial and temporal coherence properties of laser light source. All these factors significantly influence the phase measurement accuracy of QPM system and also lead to poor image quality. Recently, use of partially spatially coherent light or pseudo thermal light source (PTLS) with reduced spatial phase noise and speckle-free imaging, as compared to using a coherent laser, has been demonstrated in the field of QPM [56–59].

We have developed a QPM system using coherent laser light, partially spatially coherent laser light and low coherent light. Figure 1.9 shows the schematic diagram of experimental set-up for demonstrating the effect of speckle in QPM [57]. In this set-up we have coupled three types of light sources, i.e., highly coherent laser via single-mode fibre (SMF), and partially spatially coherent laser light using rotating diffuser and multimode fibre (MMF) bundle and temporally low coherent light

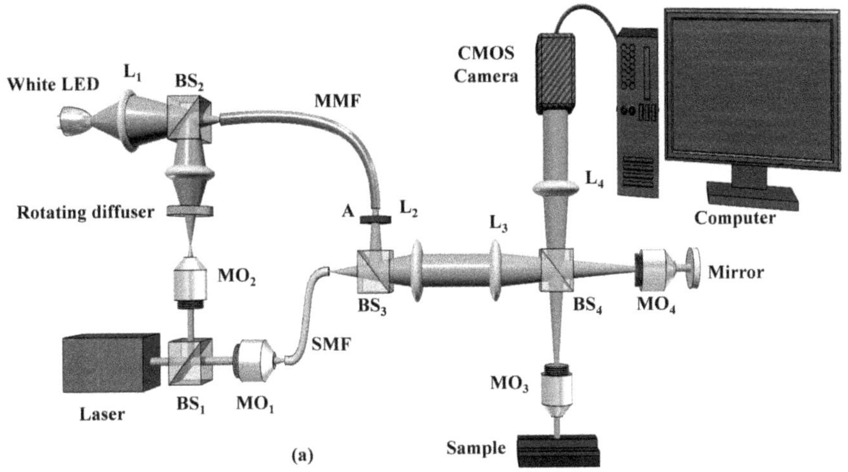

Figure 1.9. Schematic diagram of experimental set-up for demonstrating the effect of speckle in QPM. BS_{1-4}: beam splitters; MO_{1-4}: microscope objectives; L_{1-4}: lenses; RD.: rotating diffuser; A: bandpass filter; SMF: single-mode fibre; MMF: multimode fibre and computer [56].

using white LED (figure 1.9). All three light sources are combined at BS_3 and beam conditioning via lenses L_2 and L_3. The light beams are focused at the back apertures of MO_3 and MO_4 and the exiting beams are collimated, thus making the Linnik-type QPM system full field (figure 1.9). An experiment was conducted to demonstrate the effect of speckles onto the spatial phase sensitivity of the system. In figure 1.9 in place of the sample a plane mirror was placed. Interferograms were recorded using coherent light via SMF and with speckle-free partially spatially coherent light via MMF by switching them on one-by-one. To generate partially spatially coherent light the laser light passes through MO_2 and the beam passes through a rotating diffuser.

Figures 1.10(a) and (b) show interferograms recorded using coherent laser and partially coherent speckle-free laser, respectively, and figures 1.10(c) and (d) are the corresponding line profiles. It can be seen that interferogram with coherent light is imbedded with speckle noise while the interferogram with partially spatially coherent light is uniform and free from speckle noise. This non-uniform interferogram with coherent light leads to significant phase noise compared to partially coherent light. Spatial phase sensitivity was computed for both the interferograms, and figures 1.10(e) and (f) show the results for coherent and partially spatially coherent light, respectively.

It can be seen from these figures that the spatial phase noise is almost an order of magnitude less in the case of partially coherent speckle-free light source compared to coherent laser. This is the significant improvement in spatial phase noise with partially coherent light. This improvement in phase noise leads to higher accuracy in the measurement of other biophysical parameters. Further, the contrast of the interference fringes is non-uniform for coherent light (figures 1.10(a) and (c)) and uniform fringe contrast for (figures 1.10(b) and (d)).

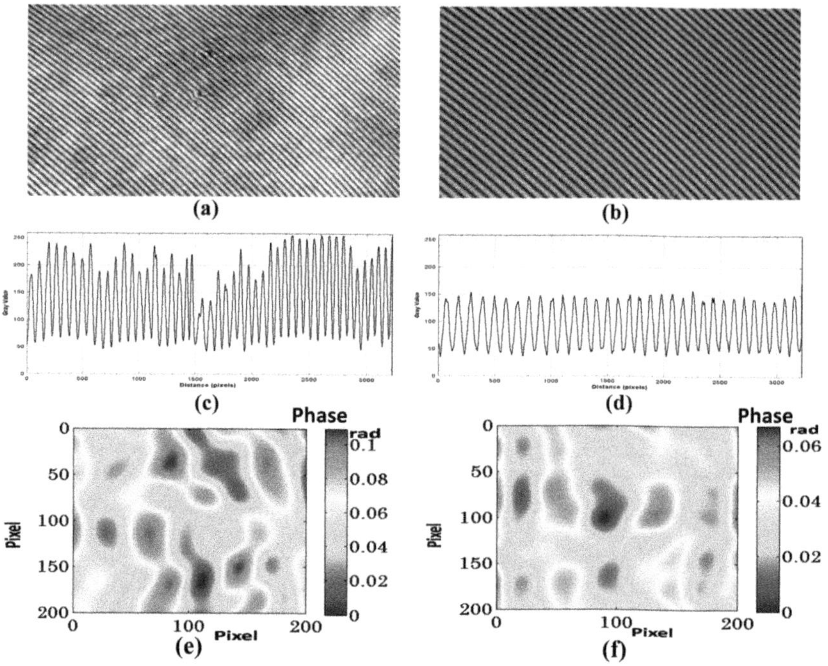

Figure 1.10. (a) and (b) interferograms, (c) and (d) corresponding line profiles with coherent laser and partially spatially coherent laser, respectively, and (e) and (f) spatial phase sensitivity, of (a) and (b).

1.6.1 Space-bandwidth product of QPM

Another important parameter in QPM is high space-bandwidth product (SBP). The high spatial phase sensitivity over the entire field of view (FOV) of interferogram is important for a single-shot QPM system. The SBP is related to the FOV and field number of the microscope objective lens. To obtain high spatial phase sensitivity one can use low temporal coherence light source, but high contrast fringes occur in only a small region or FOV. If one uses a monochromatic light source like laser whose coherence length is large, then one can obtain high contrast interference fringes over large FOV but spatial phase sensitivity is degraded [56, 59]. The SBP of a QPM system can be derived as follows. The FOV of a microscope is given by the following expression.

$$\mathrm{FOV} = \pi\, x \left(0.5\, x\, \frac{FN}{M}\right)^2 \tag{1.18}$$

where FN is the field number and is defined as 'diameter of the view field in an optical microscope and represents the diameter of the field measured in millimeters at the intermediate image plane' and M is the magnification of the microscope objective lens. The SBP of an optical microscope is the product of FOV and the area of spatial frequency information collected by the microscope and defined as follows [56, 59]:

$$\mathrm{SBP}_{\mathrm{Microscope}} = \mathrm{FOV} \times \pi(NA/\lambda)^2 \tag{1.19}$$

In QPM, the FOV of the reconstructed phase maps is limited by the high contrast interference fringe pattern FOV. Therefore, the SBP of quantitative phase microscope with suitable interferometric configuration can be computed using the following modified expression [56]:

$$SBP_{QPM} = \text{Area of interference FOV} \times \pi(NA/\lambda)^2 \tag{1.20}$$

where, $\pi(NA/\lambda)^2$is the area of the spatial frequency band, NA is the numerical aperture of the objective lens and λ is the illumination wavelength. The SBP of a QPM was realized using the experimental set-up shown in figure 1.9 for three types of light sources, i.e., temporally low coherence light (white light LED), highly coherent light source (laser) and spatially partially coherent monochromatic light source. The results are shown in figures 1.11(a)–(i). Figures 1.11(a), (d) and (g) are the schematics of three types of light source mentioned above, figures 1.11(b), (e) and (h) are the interferograms recorded using temporally low coherent (white LED), highly coherent (laser) and spatially partially coherent monochromatic light, respectively. It can be seen from these figures that SBP or the area of high contrast interferogram for QPM using low coherent light source is limited (figure 1.11(b)) while the same is large for highly coherent (figure 1.11(e)) and spatially partially

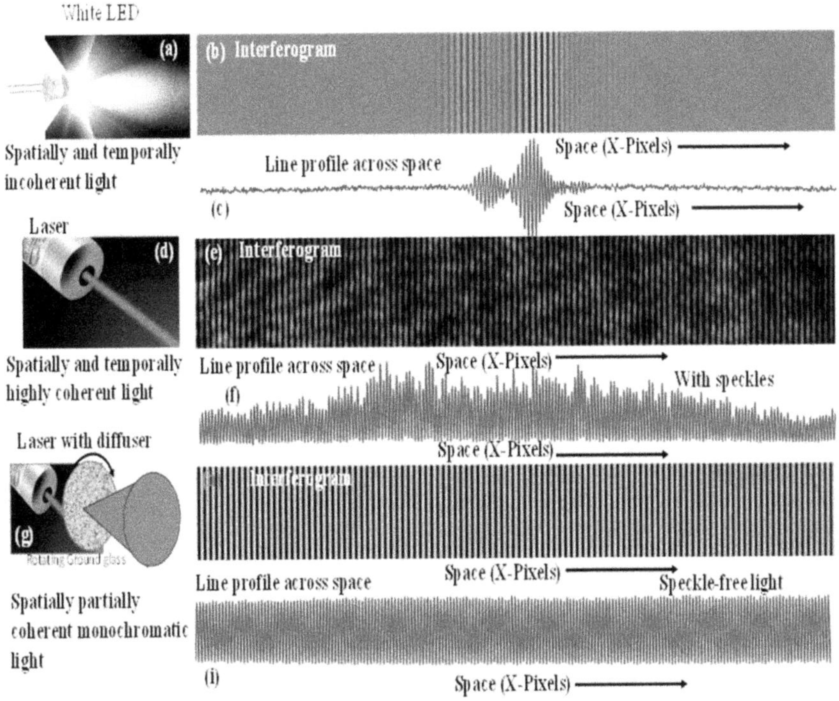

Figure 1.11. Panels (a), (d) and (g) are light sources; spatially-temporally incoherent (white LED) light, coherent (laser) and partially coherent laser. Panels (b), (e) and (h) are FOV interferograms for temporally low coherent light, coherent laser light and partially spatially coherent laser light and panels (c), (f) and (i) are corresponding line profiles [56, 59].

coherent monochromatic (figure 1.11(h)) light. Although, in the case of spatially partially coherent light the FOV or SBP for coherent light is very high, the interferogram is cluttered with speckle noise and spurious fringes (figure 1.11(e)), leading to inaccuracy in phase measurement. On the other hand, the FOV or SBP of QPM using spatially partially coherent light is large and in addition it is free- from speckle noise and spurious fringes (figure 1.11(h)), therefore, for QPM this is the most suitable light source with uniform high contrast fringes, large FOV or SBP and high spatial phase sensitivity. Figures 1.11(c), (f) and (i) are the corresponding line profiles of the interferograms throughout the FOV and they also reveal that both high contrast and high uniformity of interference fringes are obtained only with spatially partially coherent laser light. Therefore, this light source is highly recommended for QPM technique [56].

1.6.2 Effect of speckle on quantitative phase microscopy

Recently, use of partially spatially coherent light or pseudo thermal light source (PTLS) with reduced spatial phase noise and speckle-free imaging, as compared to using a coherent laser, has been demonstrated in the field of QPM. By means of using PTLS in various QPM techniques, we find that image reconstruction with reduced spatial phase noise and high accuracy in the phase and the height measurement with a factor of an order of magnitude compared to coherent light [56–65]. In addition, by means of employing PTLS in common path QPM systems, a high temporal phase stability can also be obtained along with improved spatial resolution and image sharpness [66–69]. We find that the most preferred light source for QPM, is a PTLS which gives uniform illumination, free from spurious fringes and speckled noise. In addition to this, the other remarkable properties of lasers light, i.e., its high brightness, high monochromaticity and high directionality, are fully utilized in the realization of all these techniques. Further, by means of using PTLS in QPM, the fringe visibility (degree of coherence or contrast) across the entire FOV of the interferometric image is almost constant. This enables reduced spatial phase noise, improved image sharpness and high accuracy in phase/height measurement across the entire FOV. PTLS also enables the use of both single-shot Fourier transform (FT) fringe analysis and multi-shot phase shifting interference (PSI) fringe analysis algorithms for the phase recovery. Figure 1.12 shows the typical result demonstrating the effect of speckle noise onto the imaging of QPM with coherent laser (a) and with partially spatially coherent laser (b), and (c) and (d) are the corresponding reconstructed phase maps.

1.7 Methods of generating partially spatially coherent light

There have been various methods for generating spatially partially coherent light using lasers. The most common method is laser beam passing via rotating ground glass, which is temporal diversity. Apart from this there are various other methods like spatial diversity, polarization diversity and multiple scattering [62]. In a recent review, various strategies for speckle reduction [63] have been summarised in which it has been elaborated that spatially partially coherent light is essential for laser

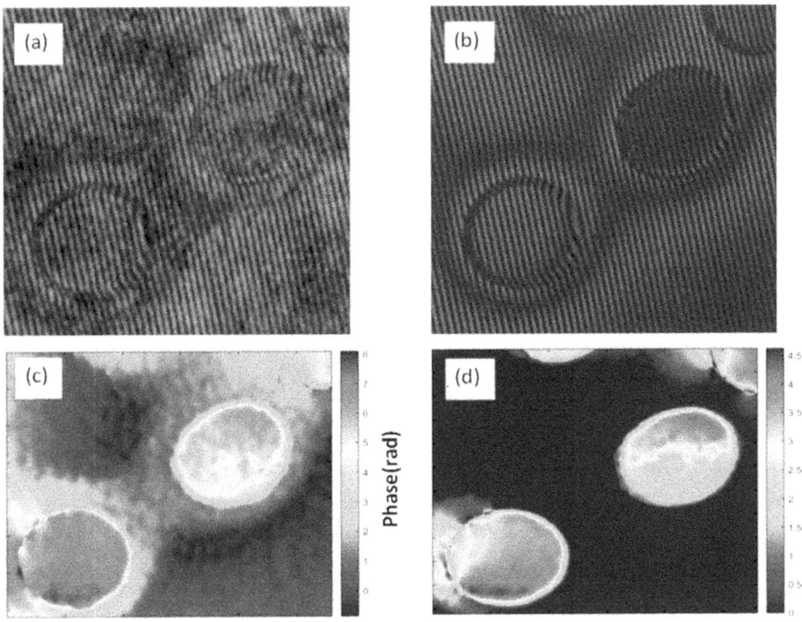

Figure 1.12. Experimental results depicting the effect of speckle in QPM imaging (a) interferometric images of red blood cells with coherent light and (b) the same with partially spatially coherent light.

projection imaging, fluorescence microscopy, digital holography and QPM. It has been demonstrated that in all the above applications while using spatially partially coherent monochromatic light one can achieve uniform illumination, improved image sharpness, improved phase sensitivity and improved resolution [62, 63]. Apart from the above methods, partially spatially coherent light can also be generated using vibrating multimode fibre bundles, using multiple scattering [70], using spatially structured and temporally varying beams and using rod-shaped waveguide diffuser and pyramidal reflective cavity without any moving parts [71–73].

References

[1] Lacey A J 1989 *Light Microscopy in Biology* (Oxford: IRL Press)

[2] Murphy D B 2002 *Fundamentals of Light Microscopy and Electronic Imaging* (New York: Wiley)

[3] Mertz J 2019 *Introduction to Optical Microscopy* (Cambridge: Cambridge University Press)

[4] Mondal P P and Diaspro A 2014 *Fundamentals of Fluorescence Microscopy: Exploring Life with Light* (Berlin: Springer)

[5] Kubitscheck U and Microscopy F 2017 *From Principles to Biological Applications* 2nd edn (New York: Wiley)

[6] Zernike F 1942 Phase contrast, a new method for the microscopic observation of transparent objects *Physica* **9** 686–98

[7] Gabor D and New A 1948 Microscopic principles *Nature* **161** 777–8

[8] Gabor D 1949 Microscopy by reconstructed wavefronts *Proc. R. Soc. (London) A* **197** 454–87

[9] Douglas Murphy B 2001 Differential interference contrast microscopy *Fundamentals of Light Microscopy and Electronic Imaging.* (New York: Wiley)

[10] Lang W 1982 *NoMarski Differential Interference-Contrast Microscopy* (Jena: Carl Zeiss)

[11] Ferraro P, Wax A and Zeev Z 2011 *Coherent Light Microscopy: Imaging and Quantitative Phase AnalysisSpringer Series in Surface Sciences* vol 46 (Cham : Springer)

[12] Popescu G 2011 *Quantitative Phase Imaging of Cells and Tissues* (New York: McGraw Hill Professional)

[13] Cuche E, Bevilacqua F and Depeursinge C 1999 Digital holography for quantitative phase contrast imaging *Opt. Lett.* **24** 291–3

[14] Kim M K 2011 *Digital Holographic Microscopy: Principles, Techniques, and Applications* (New York: Springer)

[15] Cuche E, Bevilacqua F and Depeursinge Christian 1999 Digital holography for quantitative phase-contrast imaging *Opt. Lett.* **24** 291–3

[16] Popescu G, Deflores L P, Vaughan Joshua C, Badizadegan Kamran, Iwai Hidenao, Dasari Ramachandra R and Feld Michael S 2004 Fourier phase microscopy for investigation of biological structures and dynamics *Opt. Lett.* **29** 2503–5

[17] Ikeda T, Popescu G, Dasari R R and Feld M S 2005 Hilbert phase microscopy for investigating fast dynamics in transparent systems *Opt. Lett.* **30** 1165–7

[18] Srivastava V, Anna T and Mehta D S 2012 Single shot quantitative imaging of biological cells using low coherence off-axis interferometric Hilbert phase microscopy *J. Opt.* **14** 125707

[19] Popescu T, Ikeda R R, Dasari and Feld M 2006 Diffraction phase microscopy for quantifying cell structure and dynamics *Opt. Lett.* **31** 775–7

[20] Bhaduri B, Pham H, Mir M and Popescu G 2012 Diffraction phase microscopy with white light *Opt. Lett.* **37** 1094–6

[21] Bhaduri B, Edwards C, Pham H, Zhou R, Nguyen T H, Goddard L L and Popescu G 2014 Diffraction phase microscopy: principles and applications in materials and life sciences *Adv. Opt. Photonics* **6** 57–119

[22] Pham H, Bhaduri B, Huafeng Ding and Popescu G 2012 Spectroscopic diffraction phase microscopy *Opt. Lett.* **37** 3438–40

[23] Jang Y, Jang J and Park Y 2012 Dynamic spectroscopic phase microscopy for quantifying hemoglobin concentration and dynamic membrane fluctuation in red blood cells *Opt. Express* **20** 9673–81

[24] Choi W, Fang-Yen C, Badizadegan K, Oh S, Luel N, Dasari R R and Feld M S 2007 Tomographic phase microscopy *Nat. Methods* **4** 717–9

[25] Tulsi A, Srivastava V, Mehta D S and Shakher C 2011 High-resolution full-field optical coherence microscopy using a Mirau interferometer for the quantitative imaging of biological cells *Appl. Opt.* **50** 6343

[26] Mehta Dalip Singh and Srivastava V 2012 Quantitative phase imaging of human red blood cells using phase-shifting white light interference microscopy with colour fringe analysis *Appl. Phys. Lett.* **101** 203701–3

[27] Srivastava V, Nandy S and Mehta D S 2013 High-resolution corneal topography and tomography of fish eye using wide-field white light interference microscopy *Appl. Phys. Lett.* **102** 153701

[28] Mehta D S 2018 Quantitative phase optical microscopic techniques for biomedical imaging and diagnostic applications *Proc. Natl. Acad. Sci., India, Sect. A Phys. Sci.* **88** 437–51

[29] Butola A, Ahmad A, Dubey V, Senthilkumaran P and Mehta D S 2018 Spectrally resolved laser interference microscopy *Laser Phys. Lett.* **15** 075602

[30] Popescu G, Park Y, Choi W, Dasari R, Feld M and Badizadegan K 2008 Imaging red blood cell dynamics by quantitative phase microscopy *Mol. Dis.* **41** 10–6

[31] Singh V, Joshi R, Tayal S and Mehta D S 2019 Speckle-free common-path quantitative phase imaging with high temporal phase stability using a partially spatially coherent multi-spectral light source *Laser Phys. Lett.* **16** 025601

[32] Park Y K, Depeursinge C and Popescu G 2018 Quantitative phase imaging in biomedicine *Nat. Photonics* **12** 578–89

[33] Mandel L and Wolf E 1995 *Optical Coherence and Quantum Optics* (Cambridge: Cambridge University Press)

[34] Born M and Wolf E 1999 Principles of optics: electromagnetic theory of propagation *Interference and Diffraction of Light* (Cambridge: Cambridge University Press)

[35] Saleh B E A and Teich M C 1991 *Fundamentals of Photonics* (New York: Wiley)

[36] Ghatak A 2008 *Optics* (New York: McGraw Hill Education)

[37] Drexler W and Fujimoto J G 2015 *Optical Coherence Tomography: Technology and Applications* (Berlin: Springer)

[38] Fercher A F, Drexler W, Hitzenberger C K and Lasser T 2003 Optical coherence tomography-principles and applications *Rep. Prog. Phys.* **66** 239

[39] Huang D, Swanson E A, Lin C P, Schuman J S, Stinson W G, Chang W, Hee M R, Flotte T, Gregory K and Puliafito C A 1991 Optical coherence tomography *Science* **254** 1178–81

[40] Dubois F, Requena M-L N, Minetti C, Monnom O and Istasse E 2004 Partial spatial coherence effects in digital holographic microscopy with a laser source *Appl. Opt.* **43** 1131–9

[41] Pan F, Xiao W, Liu S, Wang F J, Rong L and Li R 2011 Coherent noise reduction in digital holographic phase contrast microscopy by slightly shifting object *Opt. Express* **19** 3862–9

[42] Ahmad A, Dubey V, Singh G, Singh V and Mehta D S 2016 Quantitative phase imaging of biological cells using spatially low and temporally high coherent light source *Opt. Lett.* **41** 1554–7

[43] Edwards C, Bhaduri B, Nguyen T, Griffin B G, Pham H, Kim T, Popescu G and Goddard L L 2014 Effects of spatial coherence in diffraction phase microscopy *Opt. Express* **22** 5133–46

[44] Rodrigo J A and Alieva T 2014 Rapid quantitative phase imaging for partially coherent light microscopy *Opt. Express* **22** 13472–83

[45] Rodrigo J A and Alieva T 2014 Illumination coherence engineering and quantitative phase imaging *Opt. Lett.* **39** 5634–7

[46] Nguyen T H, Edwards C, Goddard L L and Popescu G 2014 Quantitative phase imaging with partially coherent illumination *Opt. Lett.* **39** 5511–4

[47] Nguyen T H, Edwards C, Goddard L L and Popescu G 2016 Quantitative phase imaging of weakly scattering objects using partially coherent illumination *Opt. Express* **24** 11683–93

[48] Native A and Shaked N T 2017 Compact interferometric module for full-field interferometric phase microscopy with low spatial coherence illumination *Opt. Lett.* **42** 1492–4

[49] Ghoushchi V P, Aas M, Ulusoy E and Ürey H 2017 Effect of spatial coherence of LED sources on image resolution in holographic displays *Proc. SPIE* **10126** 1–8

[50] Dubey V, Popova D, Ahmad A, Acharya G, Basnet P and Mehta D S 2019 Partially spatially coherent digital holographic microscopy and machine learning for quantitative analysis of human spermatozoa under oxidative stress condition *Sci. Rep.* **9** 1–10

[51] Choi Y, Yang T D, Lee K J and Choi W 2011 Full-field and single-shot quantitative phase microscopy using dynamic speckle illumination *Opt. Lett.* **36** 2465–8

[52] Tayal S, Usmani K, Singh V, Dubey V and Mehta D S 2019 Speckle-free quantitative phase and amplitude imaging using common-path lateral shearing interference microscope with pseudo-thermal light source illumination *Optik* **180** 991–6

[53] Mehta D S, Singh V, Tayal S, Bhatt S and Dubey V K 2020 Speckle-free quantitative phase microscopy using pseudo-thermal light source for label-free imaging of biological cells and tissues with high temporal phase stability and phase sensitivity *Proc. of SPIE, Label-free Biomedical Imaging and Sensing (LBIS)* **11251** 112511V

[54] Singh V, Tayal S and Mehta D S 2019 Simultaneous high-resolution and wide-field microscopy with improved image sharpness using pseudo-thermal light source and Fourier ptychography *Optik* **178** 1223–9

[55] Bianco V, Memmolo P, Leo M, Montresor S, Distante C, Paturzo M, Picart P, Javidi B and Ferraro P 2018 Strategies for reducing speckle noise in digital holography *Light Sci. Appl.* **7** 48

[56] Ahmad A, Dubey V, Jayakumar N, Habib A, Butola A, Nystad M, Acharya G, Basnet P, Mehta D S and Ahluwalia B S 2021 High-throughput spatial sensitive quantitative phase microscopy using low spatial and high temporal coherent illumination *Sci. Rep.* **11** 1–13

[57] Butola A *et al* 2020 High spatially sensitive quantitative phase imaging assisted with deep neural network for classification of human spermatozoa under stressed condition *Sci. Rep.* **10** 1

[58] Singh R, Dubey V, Wolfson D, Ahmad A, Butola A, Acharya G, Mehta D S, Basnet P and Ahluwalia B S 2020 Quantitative assessment of morphology and sub-cellular changes in macrophages and trophoblasts during inflammation *Biomed. Opt. Express* **11** 3733

[59] Butola A, Raosaheb Kanade S, Bhatt S, Kumar Dubey V, Kumar A, Ahmad A, Prasad D K, Senthilkumaran P, Singh Ahluwalia B and Singh Mehta D 2020 High space-bandwidth in quantitative phase imaging using partially spatially coherent digital holographic microscopy and a deep neural network *Opt. Express* **28** 36229

[60] Goodman J W 2007 *Speckle Phenomena in Optics: Theory and Applications* (Englewood, CO: Roberts & Company)

[61] Goodman J W 1975 Statistical properties of laser speckle patterns BT – laser speckle and related phenomena ed J C Dainty (Berlin: Springer) pp 9–75

[62] Dalip Singh Mehta D N, Naik R, Kumar, Singh and Takeda M 2012 Laser speckle reduction by multimode optical fiber bundle with combined temporal, spatial, and angular diversity *Appl. Opt.* **51** 1894–904

[63] Kumar V, Dubey A K, Gupta M, Singh V, Butola A and Mehta D S 2021 Speckle noise reduction strategies in laser-based projection imaging, fluorescence microscopy, and digital holography with uniform illumination, improved image sharpness, and resolution *Opt. Laser Technol.* **141** 107079

[64] Lowenthal S and Joyeux D 1971 Speckle removal by a slowly moving diffuser associated with a motionless diffuser *JOSA* **61** 847–51

[65] Kubota S and Goodman J W 2010 Very efficient speckle contrast reduction realized by moving diffuser device *Appl. Opt.* **49** 4385–91

[66] Tayal S, Singh V, Kaur T, Singh N and Mehta D S 2022 Multimodal biomicroscopic system for the characterization of cells with high spatial phase sensitivity and sub-pixel accuracy *J. Biophoton.* **15** e202100258

[67] Butola Ankit *et al* 2022 Multimodal on-chip nanoscopy and quantitative phase imaging reveals the nanoscale morphology of liver sinusoidal endothelial cells *Proc. Natl Acad. Sci.* **118** e2115323118

[68] Tayal S, Singh V, Kaur T, Singh N and Mehta D S 2020 Simultaneous fluorescence and quantitative phase imaging of MG63 osteosarcoma cells to monitor morphological changes with time using partially spatially coherent light source *Methods Appl. Fluoresc.* **8** 035004
[69] Ahmad A, Dubey V, Butola A, Tinguely J C, Ahluwalia B S and Mehta D S 2020 Sub-nanometer height sensitivity by phase shifting interference microscopy under environmental fluctuations *Opt. Express* **28** 9340–58
[70] Redding B, Allen G, Dufresne E R and Cao H 2013 Low-loss high-speed speckle reduction using a colloidal dispersion *Appl. Opt.* **52** 1168–72
[71] Kumar V, Usmani K, Singh V, Dubey A K, Gupta M and Mehta D S 2020 Laser speckle reduction using spatially structured and temporally varying beams using double diffractive optical elements *Laser Phys. Lett.* **17** 036003
[72] Kumar V, Gupta M, Dubey A K, Tayal S, Singh V and Mehta D S 2020 Design and development of laser speckle reduction device using waveguide diffuser and pyramidal cavity for projection imaging *J. Opt.* **22** 115601–10
[73] Kumar V, Dubey A K, Gupta M and Mehta D S 2022 Speckle free efficient light engine for high power laser projectors and automobile headlamps *Opt. Laser Technol.* **152** 108145

Chapter 2

Partially spatially coherent off-axis quantitative phase microscopy

Off-axis holography is a non-invasive and label-free optical technique to extract complete information (amplitude and phase) of specimens. In this chapter, we highlight the reading and writing of a conventional off-axis hologram. In addition, digital off-axis holography and the associated reconstruction approaches to retrieve phase map of the object are also shown. Importantly, high spatial coherence of the light source reduces the quality of phase reconstruction and therefore, the advantage of partially spatially coherent source in off-axis holography is presented. Finally, extraction of quantitative features for the interpretation of phase value is shown, which can be useful for disease diagnosis and classification problems.

2.1 Off-axis holographic phase microscopy

Holography was first introduced by Dennis Gabor in 1948 [1] for which he was awarded the Nobel prize in Physics in 1971. Holographic imaging indicates the idea of recording the complete information, i.e., amplitude and phase of the light waves scattered by specimens. Since all the detectors are sensitive to intensity of light only, it is inevitable to encode the phase information into variations of intensity to extract the complete information of the object. The most common way of performing it is to interfere the reference wave and scattered wave by an object, both propagating different optical paths, and is known as off-axis holography, first proposed by Leith and Upatnieks [2]. Conventional off-axis holography is a two-step process: (a) recording of the hologram; and (b) reconstruction of the hologram. Interestingly, quantitative phase imaging (QPI) fundamentally exploits the idea of off-axis holography to extract the path length shift associated with the object. The path length shift at each point contains the information of refractive index and geometrical thickness of the sample and thus has gained remarkable popularity in many biological applications.

doi:10.1088/978-0-7503-3987-2ch2

2.1.1 Conventional (Leith and Upatnieks') holography

In 1962, Leith and Upatnieks [2] proposed off-axis holography by introducing an off-axis reference beam to separate the real and twin image and zero order diffraction terms. The principle of recording and reading of off-axis hologram is depicted in figure 2.1. To record the off-axis hologram, the object is illuminated by a plane wave E_0. The field is transmitting through the object and propagating to the distance z can be written as [3, 4]:

$$E_s(x, y) = E(x, y) \otimes e^{\frac{ik_0(x^2+y^2)}{2z}} \tag{2.1}$$

where k_0 is a wavevector and z is a propagation distance. To record the hologram, reference field is a plane wave at some angle, say θ, with respect to the z-axis. The total field can be represented mathematically as [4]:

$$E_t(x, y) = E_s(x, y) + | E_r | e^{i (xk_0 \sin\theta)} \tag{2.2}$$

z components of the reference field provide a constant phase shift, and are therefore not considered in equation (2.2). The transmission function i.e., $t(x,y)$ of the hologram can be written as:

$$\begin{aligned} t(x, y) = {} & | E_s(x, y) |^2 + | E_r |^2 + E_s(x, y) | E_r | e^{-i (xk_0 \sin\theta)} \\ & + E_s(x, y)^* | E_r | e^{i (xk_0 \sin\theta)} \end{aligned} \tag{2.3}$$

On the other hand, the same reference field E_r is used to illuminate the hologram at the same angle to reconstruct it. Mathematically, illuminating the hologram (equation (2.3)) with a plane wave at angle θ can be written as:

$$\begin{aligned} E_t(x, y) = {} & t(x, y) | E_r | e^{i (xk_0 \sin\theta)} = | E_s(x, y) |^2 | E_r | \\ & e^{i (xk_0 \sin\theta)} + | E_r |^2 + E_s(x, y) | E_r |^2 + E_s(x, y)^* | E_r |^2 e^{i 2(xk_0 \sin\theta)} \end{aligned} \tag{2.4}$$

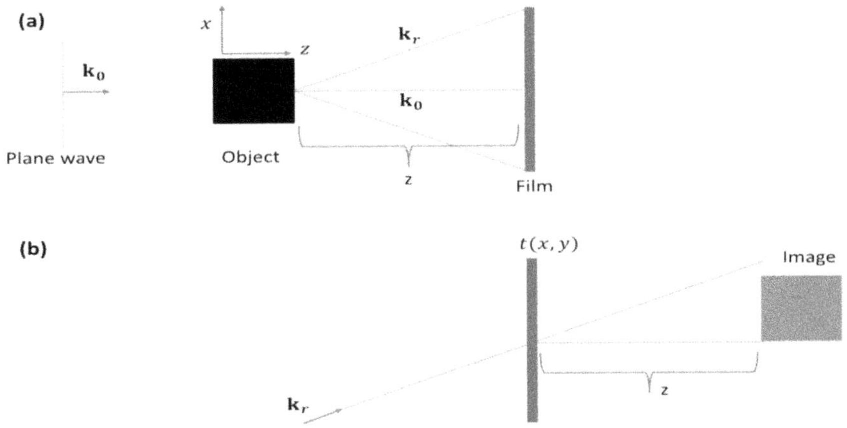

Figure 2.1. Schematic diagram of recording (a) and reading (b) conventional (Leith and Upatnieks') hologram.

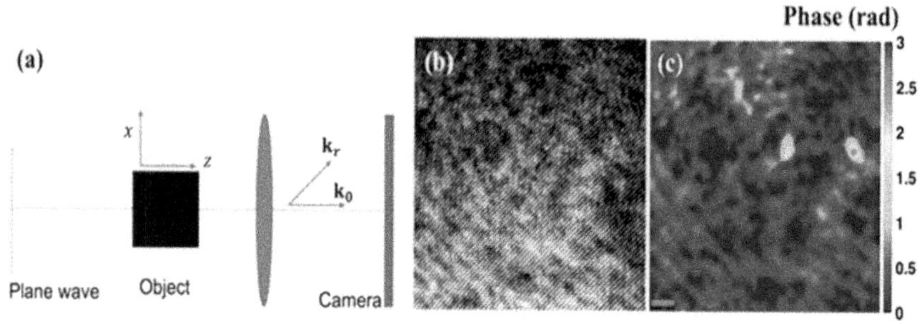

Figure 2.2. Schematic diagram of recording (a) off-axis image plane hologram. (b) Recorded hologram and the (c) reconstructed phase image of the object. Scale bar: 5 μm

The first two terms in equation (2.4) represent the field propagation. The third term represents the complex sample field along the z-direction while the fourth term is the virtual image of the object.

2.1.2 Digital holography

Since conventional off-axis (Leith and Upatnieks') holography [2] was first introduced in 1962, it has been developed both computationally and experimentally to retrieve the real-time image of the specimens. During the last several decades digital holography has become more popular due to the availability of solid state electronics based area detectors and computers [5]. Digital holography simply replaces the film-based writing of the hologram with digital camera, CCD or CMOS (figure 2.2(a)) and a variety of computational reconstruction algorithms are used to retrieve the information of the object [5, 6]. Generally, to record a digital hologram usually a highly coherent laser source is used. The recorded hologram is full of coherence noise, as shown in figure 2.2(b). The intensity distribution at the camera plane due to the interference between sample and reference beam can be written as [6]:

$$I(x, y) = a(x, y) + b(x, y)\cos\left[2\pi i(f_x x + f_y y) + \phi(x, y)\right] \tag{2.5}$$

$a(x,y)$ is the background (DC) and $b(x,y)$ is known as modulation term. $\phi(x,y)$ is a spatially varying phase which contains information of the sample and f_x, f_y are the spatial frequencies of hologram. In addition, intensity modulation can be rewritten as [5]:

$$\begin{aligned} I(x, y) = a(x, y) + c(x, y)\exp[2\pi i(f_x x + f_y y)] \\ + c^*(x, y)\exp[-2\pi i(f_x x + f_y y)] \end{aligned} \tag{2.6}$$

where

$$c(x, y) = B(x, y)\exp(i\phi(x, y)) \tag{2.7}$$

Different techniques have been proposed to reconstruct the phase $\phi(x,y)$ of the object. For example, Fourier transform (FT) based approach is widely adopted to reconstruct the interferogram (figure 2.2(b)) and filter out the complex information, as shown in figure 2.2(c). It is important to note that the Fresnel propagation transformation is used if the object is placed at an arbitrary distance from the lens.

2.1.2.1 Fourier transform based phase reconstruction technique

The FT [7] reconstruction technique is widely used for the single-shot phase reconstruction of the hologram. In FT based reconstruction technique, 2D Fourier transform operation is performed to separate the DC and real part of the complex correlation function. Fourier transform of the intensity modulation equation (2.6) can be written as:

$$FI(\xi_x, \xi_y) = Fa(\xi_x, \xi_y) + Fc(\xi_x - f_x, \xi_y - f_y) + Fc^*\left[-(\xi_x + f_x), -(\xi_y + f_y)\right] \tag{2.8}$$

where $Fa(\xi_x, \xi_y)$ represents DC i.e., background term at the center of Fourier plane. $Fc(\xi_x - f_x, \xi_y - f_y)$ contains object information at $(+f_x, +f_y)$ coordinate known as +1 term. Similarly, $Fc^*\left[-(\xi_x + f_x), -(\xi_y + f_y)\right]$ is known as −1 order term at $(-f_x, -f_y)$ coordinate and contains complex conjugate information of the object. Applying Fourier filtering of −1 and zero order, equation (2.8) can be written as:

$$FI(\xi_x, \xi_y) = Fc(\xi_x - f_x, \xi_y - f_y) \tag{2.9}$$

The filtered spectrum is, first, shifted at the origin and then inverse Fourier transformed to extract the complex signal $c(x, y)$, subsequently the wrapped phase map from the following expression:

$$\phi(x, y) = \tan^{-1}\left[\frac{\mathrm{Im}(c(x, y))}{\mathrm{Re}(c(x, y))}\right] \tag{2.10}$$

where Im and Re are the imaginary and real part of the complex signal, respectively. In addition, a reconstructed wrapped phase map lies between −π to +π which can be corrected by unwrapping an algorithm such as in Goldstein *et al* [8] and minimum Lp norm method [9]. However, these algorithms suffer from some streaks at the edges of samples, especially in the case of a thick sample. Therefore, unwrapping of the phase map was recently proposed by using transport of intensity (TIE) equations [10, 11]. It has been shown that TIE unwrapping is faster compared to other algorithms and performs better for a thicker sample [12]. In TIE phase unwrapping, constant amplitude, and phase $\phi(x, y)$ are used to create a complex field. Mathematically:

$$c_o(x, y; 0) = \exp(i\phi(x, y)) \tag{2.11}$$

The field $c_o(x, y; 0)$ is propagated to $\pm z$, i.e., a closely spaced plane to derive the longitudinal intensity derivative:

$$c_o(x, y; \pm z) = c_o(x, y; 0) * s(x, y; \pm z) \tag{2.12}$$

Equation (2.12) represents the convolution between impulse response and complex field for free-space propagation where $s(x, y; z)$ is the impulse response and can be derived as [13]:

$$s(x, y; z) = \frac{e^{ikR}}{2\pi R}\left(-ik + \frac{1}{R}\right)\frac{z}{R} \tag{2.13}$$

The angular spectrum transfer function for free-space propagation can be used to calculate the impulse response using the following expression [3]:

$$S\left(f_x, f_y; z\right) = F[s(x, y; z)] = \exp\left[iz\sqrt{k^2 - 4\pi^2\left(f_x^2 + f_y^2\right)}\right] \tag{2.14}$$

Using equation (2.12), equation (2.12) and central difference relation, longitudinal intensity derivative can be estimated as [14]:

$$\frac{\partial I}{\partial z} = \frac{\mid c_o(x,\ y;\ \Delta z)\mid^2 - \mid c_o(x,\ y;\ -\Delta z)\mid^2}{2\Delta z} \tag{2.15}$$

Finally, applying inverse Laplacian transform in longitudinal intensity derivate, the unwrapped phase can be written as [14]

$$\phi(x, y; z) = \frac{-k}{I}\nabla_{x,y}^{-2}\left(\frac{\partial I}{\partial z}\right) \tag{2.16}$$

Equation (2.16) shows unwrapped phase of an object. Since the TIE unwrapping is implemented with FFT algorithm, the timing performance of the combined computation is better than any other phase retrieving algorithm. Further, the retrieved unwrapped phase map can be further used to calculate the corresponding height map.

2.1.2.2 Phase shifting interferometry-based reconstruction technique

Phase shifting interferometry (PSI) is a well-established technique to determine the phase map of the light interrogated with the specimen [15, 16]. In contrast to the FT based reconstruction method, PSI preserves lateral resolution with a cost of low acquisition rate [17–19]. PSI is used to implement the precise phase steps between sample and reference beams, i.e., $\delta\phi$ and the interference expression can be written as;

$$I\ (x, y) = I_s(x, y) + I_r(x, y) + 2\sqrt{I_s I_r}\,\cos[\phi(x, y) + \delta\phi] \tag{2.17}$$

where $\delta\phi$ is the known phase step introduced between sample and reference field. Various phase shifting algorithms (three-step, four-step, five-step and so on) have been proposed to compute the phase map $\phi(x, y)$ of the object. Among all these methods the five-step phase shifting interferometry is widely adopted because of moderate acquisition rate and less phase error [20]. Consider a controlled phase shift $\delta\phi$ introduced in the reference arm, five-step phase shifted interferograms can be written as

$$I_1(x, y) = I_s(x, y) + I_r(x, y) + 2\sqrt{I_s I_r}\cos[\phi(x, y) - \pi] \tag{2.18}$$

$$I_2(x, y) = I_s(x, y) + I_r(x, y) + 2\sqrt{I_sI_r}\cos\left[\phi(x, y) - \frac{\pi}{2}\right] \tag{2.19}$$

$$I_3(x, y) = I_s(x, y) + I_r(x, y) + 2\sqrt{I_sI_r}\cos[\phi(x, y)] \tag{2.20}$$

$$I_4(x, y) = I_s(x, y) + I_r(x, y) + 2\sqrt{I_sI_r}\cos\left[\phi(x, y) + \frac{\pi}{2}\right] \tag{2.21}$$

$$I_5(x, y) = I_s(x, y) + I_r(x, y) + 2\sqrt{I_sI_r}\cos[\phi(x, y) + \pi] \tag{2.22}$$

These equations can be utilized to calculate phase map $\phi(x,y)$ of the light field interrogated with the object using an expression:

$$\phi(x, y) = \tan^{-1}\left[\frac{2(I_4(x, y) - I_2(x, y))}{I_1(x, y) - 2I_3(x, y) + I_5(x, y)}\right] \tag{2.23}$$

The reconstructed phase map is related to the refractive index and geometrical thickness of the samples by the following expression:

$$\phi(x, y) = \frac{2\pi}{\lambda}[n_c(x, y) - n_m(x, y)] \times 2d(x, y) \tag{2.24}$$

where $n_c(x, y)$ and $n_m(x, y)$ are the refractive indices of sample and surrounding medium, respectively, and $d(x,y)$ is the cell thickness. Though phase shifting interferometry offers high spatial resolution but is less preferred because of slow acquisition rate and thus cannot be applicable for live cells imaging. Nevertheless, one possible solution is that if all five-step phase shifted interferograms are recorded quickly using a high speed camera operating much higher speed compared to the dynamic fluctuations of a cell, then phase shifting interferometry is the best. Another possible solution is that the parallel four-step phase shifting along with single-shot recording can also be used for dynamic samples.

2.2 Partially spatially coherent off-axis quantitative phase microscopy

Spatial coherence is the correlation of optical fields at two different spatial locations (transverse or longitudinal direction) of the beam propagation [21, 22]. Interestingly, spectral bandwidth of the light source describes temporal coherence function and spatial coherence function is decided by angular frequency spectrum, i.e., number of spatial frequencies contained in the light source [21, 22]. A highly spatially and temporally coherent source, such as, He–Ne laser degrades the image quality due to coherent noise and the formation of spurious interference fringes due to multiple reflections from the optical components and from multiple layers of biological specimens and reduces phase and height measurement accuracy of digital holography microscopy (DHM) systems.

Comparison between a highly spatially/temporally coherent He–Ne laser and a partially spatially coherent He–Ne laser is depicted in figure 2.3. The interferometric images of the sperm cells are recorded using a partially spatially coherent light based

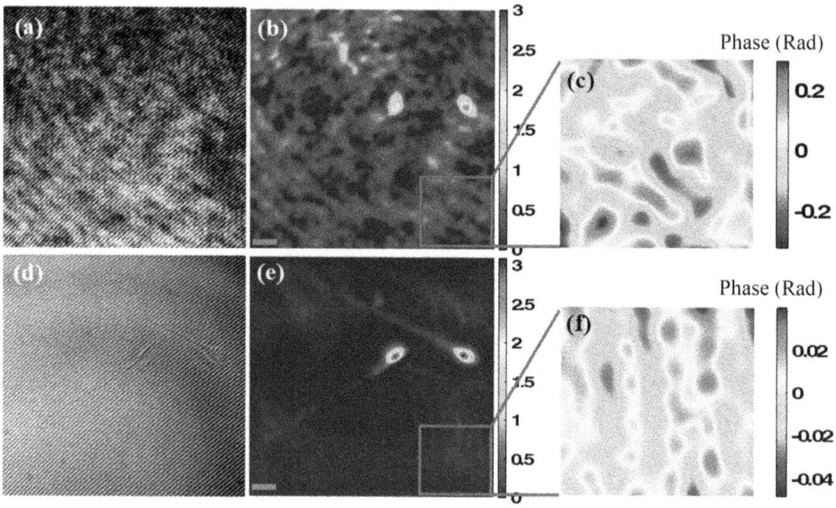

Figure 2.3. Panels (a), (b) and (c) are the interferogram, 2D-phase map, and spatial phase sensitivity with sperm cells using coherent laser light, and panels (d), (e) and (f) are interferogram, 2D-phase map, and spatial phase sensitivity with sperm cells using partially spatially coherent light. Color bar represents the phase map in radian. Reproduced from [23], copyright (2020) The Authors, with permission from Springer.

digital holography microscopy (PSC-DHM) system. Figure 2.3 represents the phase reconstruction comparison between the highly coherent laser and partially spatially coherent light, i.e., pseudo thermal light source. It can be seen that speckle pattern degrades the contrast and uniformity of the interferogram (see figure 2.3(a)) significantly and therefore, poor spatial phase sensitivity (±200 mrad) can be seen using a coherent laser (figure 2.3(c)). High spatial phase sensitivity over the entire field of view is important for the accurate measurement of the phase map of the light field interacting with the object, therefore, the smaller the value of phase fluctuations across the field of view, the higher is the accuracy in phase measurement [23, 24]. Therefore, the thinnest part of the sperm cells, i.e., the tail part cannot be clearly seen with direct laser illumination. Additionally, spatial phase sensitivity in the case of partially spatially coherent light illumination (figures 2.3(d)–(f)) is found approximately 10× higher than the coherent source and therefore phase reconstruction of the tail part of the cell can be clearly visible, as in figure 2.3(e). Different off-axis optical configurations have been proposed in recent years to reconstruct the phase, i.e., combined information of refractive index and thickness of the specimens. In the following part we will focus on different off-axis digital holographic configurations.

2.2.1 Linnik interferometer based optical configuration

A partially spatially coherent Linnik-type system for quantitative phase imaging (QPI) is depicted in figure 2.4. To reduce the spatial coherence, the laser beam was first focused by an objective lens (MO_1) onto the rotating ground glass diffuser which was placed in the front focal plane of the MO_1. The rotating ground glass diffuser scatters the laser light randomly into multiple directions and is collected by a multi-multimode fiber

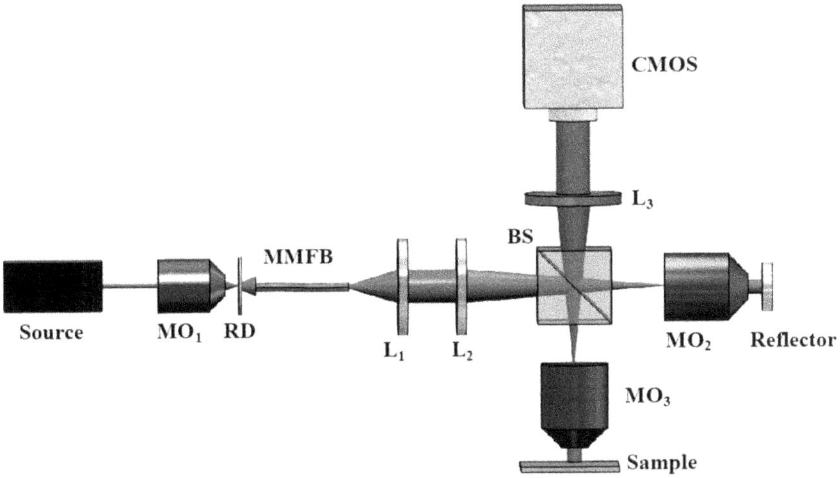

Figure 2.4. Partially spatially coherent Linnik-type DHM system. Reproduced from [23], copyright (2020) The Authors, with permission from Springer.

bundle (MMFB). The light rays are multiply scattered inside the MMFB and the output field consists of temporally high but spatially partially coherent light which acts as an extended light source. The extended light source is coupled in the Linnik type interferometer. A partially spatially coherent beam is collimated by using lens L_1 followed by L_2, to focus the light into the back focal plane of reference and objective lenses. A beam splitter (BS) is placed in between the path to divide the focused light beam into sample and reference arms. The output laser beam from the sample and reference arms is nearly collimated and the interference takes place from which the phase map can be extracted of the sample. The spatially partially coherent light beam is backscattered from the specimen and the reference mirror interferes at the BS plane and the interferogram is finally projected into the camera sensor by using a tube lens L_3.

2.2.2 Mach–Zehnder interferometer based optical configuration

In most of the DHMs a highly coherent laser is used as light source. The laser beam is first spatially filtered, expanded, and then collimated. This can be done using a microscope objective (MO) lens followed by a pinhole (5–10 μm) kept at the focus of MO and a collimating lens kept at front focal plane. This complete arrangement is necessary while using the laser as a light source for each interferometric or holographic experiment including quantitative phase microscopy (QPM). The spatially filtered, expanded and collimated laser beam is free from noise. Figure 2.5 depicts the Mach–Zehnder interferometer based off-axis DHM system [25, 26]. In this geometry, the interferogram is recorded in the image plane and preferred to image optically thin samples such as cells. In a Mach–Zehnder interferometer, a spatially filtered laser beam is split into object beam and reference beam using a BS. The object beam illuminates the sample, and the scattered field is collected by the microscopic objective (MO) lens. The tube lens is used to transfer the image at the camera, i.e., charge coupled device (CCD) plane. The reference beam is tilted to generate the high fringe density. But the use of

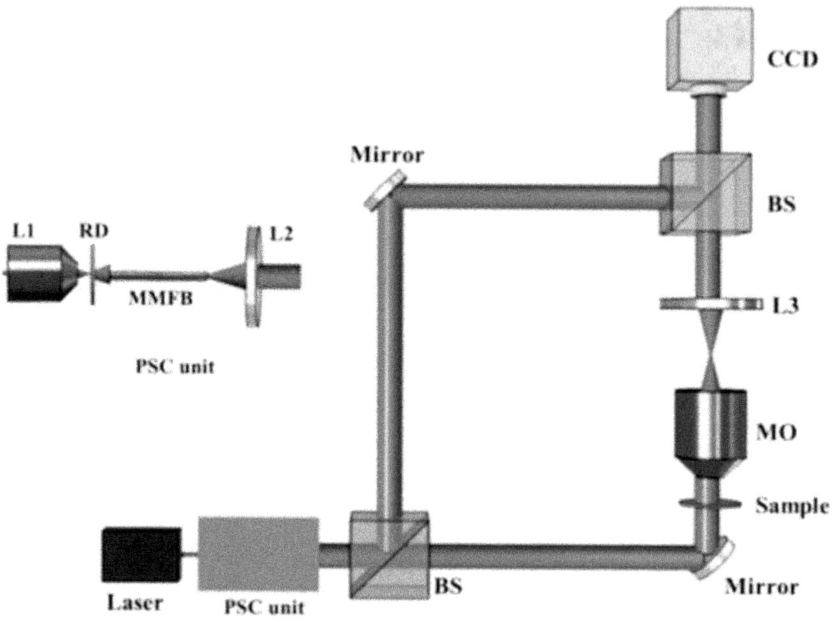

Figure 2.5. Off-axis digital holography microscopy based on Mach–Zehnder interferometer system. L: lens; BS: beam splitter; MO: microscopic objective lens; CCD: charge coupled device.

coherent laser is only suitable for mirror-like objects, on the other hand while using realistic objects such as rough surfaces and any biological sample, the hologram or interferogram is imbedded with speckle noise and spurious fringes. This problem can be solved using spatially partially coherent light. The inset of figure 2.5 shows the unit for generating spatially partially coherent light. But while using spatially partially coherent light in a Mach–Zehnder interferometer the temporal coherence length of laser must be large, therefore, interference fringes can be obtained easily.

Nativ and Shaked [27] developed a compact interferometric system using a partially spatially coherent light source which can generate high visibility interferograms over the entire field of view of the camera. The developed compact interferometer is easy to align and stable and can be attached at the output of an optical microscope illuminated with a low spatial coherence light source, thus allowing QPI with a low degree of spatial noise. Figure 2.6 shows the comparison of the imaging properties and the interference visibility.

Choi *et al* [28] developed an off-axis QPM using short spatial coherence length light source. They developed an off-axis dynamic speckle interferometric microscopic (ODSIM) set-up using a rotating ground glass diffuser and a Mach–Zehnder interferometric set-up. The method was applied to image the dynamics of small intracellular particles in live biological cells. Figures 2.7(a), (c) and (e) are raw interferogram, processed quantitative phase image, and numerically simulated differential interference contrast (DIC) image under the coherent light illumination, (b), (d) and (f) are same as (a), (c), and (e) using partially spatially coherent light. The phase

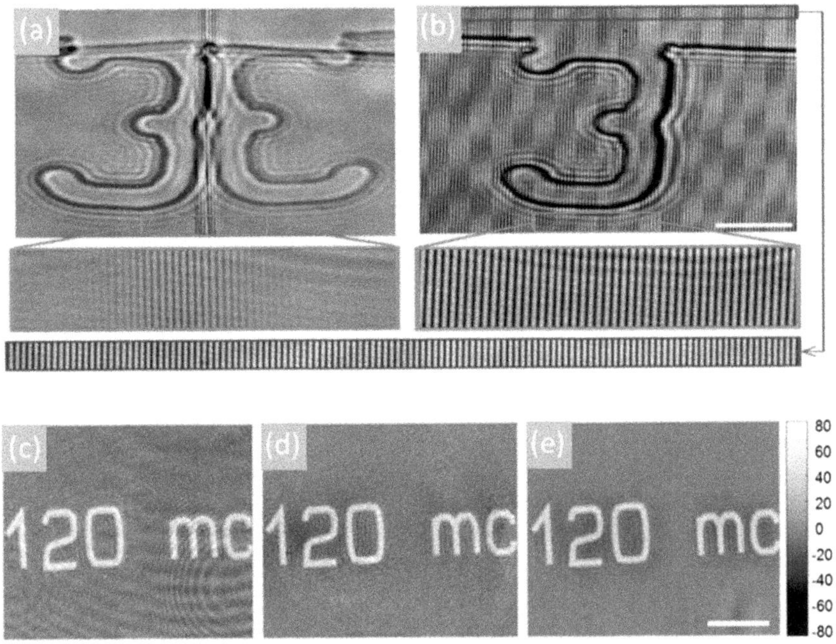

Figure 2.6. Comparison of the imaging properties and the interference visibility between (a) the pre design and (b) the post design, both under low spatial coherence illumination, panels (c)–(e) show OPD measurements with three degrees of spatial coherence of letters etched by an electron beam on a silicon wafer, demonstrating the decrease of spatial noise obtained when the spatial coherence of the illumination decreases. (c) High degree of spatial coherence; $\rho_c = 28$ μm, taken using the previous module. (d) Medium degree of spatial coherence; $\rho_c = 7$ μm, taken using the proposed PC τ module. (e) Low degree of spatial coherence; $\rho_c = 4$ μm, taken using the proposed PC τ module. The white scale bar represents 10 μm on the sample. Reprinted with permission from [27]. Copyright 2017 Optical Society of America.

sensitivity was enhanced more than a factor of 10 in comparison with coherent illumination. Significant improvement in the accuracy of phase measurement was achieved using partially spatially coherent light compared to coherent light.

2.3 Image characteristics

The main quality of the QPI system is defined by four factors, i.e., transverse resolution, acquisition rate, spatial phase sensitivity and temporal phase sensitivity.

2.3.1 Resolution

The resolution of the QPI system is defined by minimum distance between two closely spaced points that are considered resolved. The resolution in a QPI system mainly depends on the numerical aperture (NA) of the microscopy objective lens and associated phase reconstruction technique. Nonetheless, a few computational reconstruction techniques have been proposed in recent years that utilize full resolution of the system and therefore, the transverse resolution mainly depends on the NA of the lens.

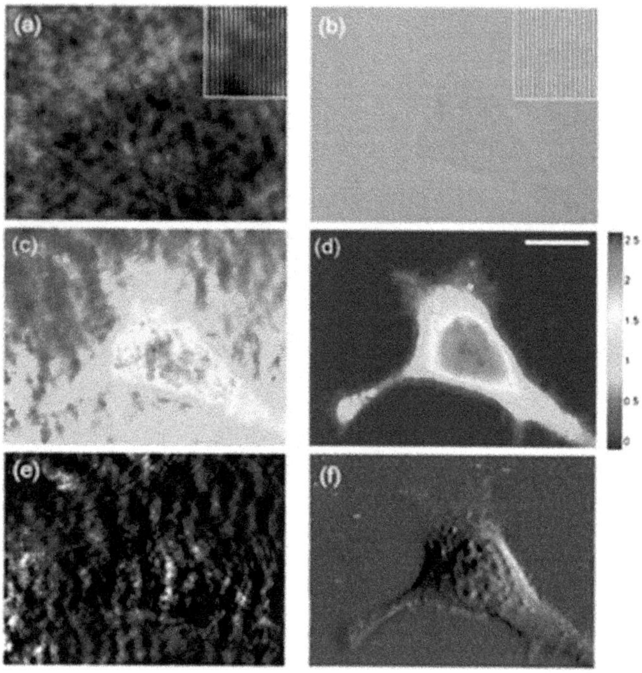

Figure 2.7. QPI of a live cell using off-axis dynamic speckle interferometric microscopy: (a) raw interference image, (c) processed quantitative phase image, and (e) numerically simulated DIC image for the coherent illumination. (b), (d), and (f) Same as (a), (c), and (e) but with dynamic speckle illumination. The insets in (a) and (b) are zoom-in images at the background by three folds. Scale bar, 10 μm. Color bar, phase in radians. Reprinted with permission from [28]. Copyright 2011 Optical Society of America.

In an imaging system, the output image $I(r)$ of an object $O(r)$ is defined as [29]:

$$I(r) = O(r) \otimes h(r) \tag{2.25}$$

where $h(r)$ represents the impulse response and generally called point spread function (PSF) of the microscopy system. Fourier transform of equation (2.17) can be written as

$$I(k) = O(k) \times h(k) \tag{2.26}$$

$O(k)$ is the Fourier transform of $O(r)$ and $h(k)$ is Fourier transform pair of $h(r)$ known as transfer function (TF). The image formation mainly depends on TF and PSF of the system. A high resolution requires narrow PSF which means broader TF of the system.

2.3.2 Signal-to-noise ratio

Signal-to-noise ratio in a microscopy system is defined by the variation of the signal around the average signal for say N measurements, mathematically [14, 29]:

$$\sigma^2 = \frac{1}{N}\sum_{i=1}^{N}(x_i - \hat{x})^2 \tag{2.27}$$

x_i represents the captured signal and x is defined as:

$$\hat{x} = \frac{1}{N}\sum_i x_i \tag{2.28}$$

The noise can be characterized both spatially and temporally. Spatial noise can be calculated if measurements are performed over N pixels of the same image. On the other hand, if the noise is calculated for the same point at N different times then it can be characterized and is temporal noise. The concept is very important and widely used in QPI to calculate temporal phase sensitivity and spatial phase sensitivity.

Spatial phase sensitivity can be obtained by means of calculating the phase image without any sample. The standard deviation in the phase image without any sample can be defined mathematically [29, 30]:

$$\sigma_r = \sqrt{\langle[\delta\phi(x, y) - \langle\delta\phi(x, y)\rangle]^2\rangle} \tag{2.29}$$

Similarly, temporal phase sensitivity can be calculated over successive measurements without any sample [29].

$$\sigma_t = \sqrt{\langle[\delta\phi(t) - \langle\delta\phi(t)\rangle]^2\rangle} \tag{2.30}$$

$\delta\phi(x,y)$ and $\delta\phi(t)$ are defined as spatial and temporal phase fluctuation, respectively. Note that white light source offers the maximum spatial phase sensitivity while a common-path interferometric system provides best temporal phase sensitivity in the QPM system.

2.4 Extraction of quantitative information using phase map

The analysis of extracted phase of the object is crucial for the image-based computer-aided diagnosis (CAD) to offer best accuracy in early-stage disease detection. Various morphological features such as surface area, volume, area-to-volume ratio (S/V), sphericity and optical thickness can be obtained from the reconstructed phase map. Surface area can be calculated using Monge parameterization defined in table 2.1, where g_x and g_y are the gradients along the x and y directions, respectively. In addition, dx and dy are the calibrated pixel width along x and y directions, respectively. Volume of the cell in the phase image can be calculated by first calculating optical thickness of the cell.

Table 2.1 shows the relation between optical thickness (OT) in the reflection mode to the phase value of the object. The measured OT is utilized to measure the volume of sperm head and can be calculated by integrating OT over projected area. S/V and sphericity can be calculated from the known surface area and volume of the cell. Other statistical parameters such as mean, variance, energy, entropy, skewness, and kurtosis are used in recent studies for healthy/unhealthy cell detection. These features can be calculated from the mathematical formulae mentioned in table 2.1.

Table 2.1. Extraction of morphological and texture features from the phase map [23, 31]

Feature	Formula
Surface area (*S*)	$\iint \sqrt{1 + g_x + g_y}\, dxdy$
Volume (*V*)	$\iint OT(x, y)dxdy$
S/*V*	S/V
Sphericity	$\left(4.84 \times V^{\frac{2}{3}}\right)/_S$
Optical thickness (*OT*)	$\phi(x, y) \times \lambda/4\pi$
Mean(μ)	$\frac{1}{N}\frac{1}{M}\sum_{i=1}^{N}\sum_{j=1}^{M}\phi_{ij}$
Variance (σ^2)	$\frac{1}{N}\sum_{i=1}^{N}\sum_{j=1}^{M}\left(\phi_{ij} - \mu\right)^2$
Energy	$\sum_{i,\, j=0}^{N-1}\left(\phi_{ij}\right)^2$
Entropy	$-\sum_{i=1}^{N} p(\phi_i) \log_2 p(\phi_i)$
Skewness	$\frac{1}{N}\sum_{i=1}^{N}\sum_{j=1}^{M}\left[\frac{(\phi_{ij} - \mu)}{\sigma}\right]^3$
Kurtosis	$\frac{1}{N}\sum_{i=1}^{N}\sum_{j=1}^{M}\left[\frac{(\phi_{ij} - \mu)}{\sigma}\right]^4 - 3$

In summary, off-axis QPM offers unique advantage to the label-free microscopy community. Optical phase value of the specimens can be retrieved using different experimental and computational approaches. Interpretation of the phase value has proven to offer various parameters for studying live cell dynamics, dry mass, sphericity, and density change. These parameters can be used in disease identification and, therefore, QPM will be a potential tool for disease diagnosis. Commercialization and translating discoveries to clinical applications will be the key for future developments.

References

[1] Gabor D and New A 1948 Microscopic principles *Nature* **161** 777–8

[2] Leith E N and Upatnieks J 1962 Reconstructed wavefronts and communication theory *JOSA* **52** 1123–30

[3] Goodman J W 2005 *Introduction to Fourier Optics* (Englewood, CO: Roberts & Company)

[4] Saleh B E and Teich M C 2019 *Fundamentals of Photonics* (New York: Wiley)

[5] Poon T-C 2006 *Digital Holography and Three-Dimensional Display: Principles and Applications* (Berlin: Springer Science & Business Media)

[6] Poon T-C and Liu J-P 2014 *Introduction to Modern Digital Holography: With MATLAB* (Cambridge: Cambridge University Press)

[7] Takeda M, Ina H and Kobayashi S 1982 Fourier-transform method of fringe-pattern analysis for computer-based topography and interferometry *JOSA* **72** 156–60

[8] Goldstein R M, Zebker H A and Werner C L 1988 Satellite radar interferometry: two-dimensional phase unwrapping *Radio Sci.* **23** 713–20

[9] Ghiglia D C and Romero L A 1996 Minimum Lp-norm two-dimensional phase unwrapping *J. Opt. Soc. Am. A* **13** 1999–2013

[10] Paganin D and Nugent K A 1998 Noninterferometric phase imaging with partially coherent light *Phys. Rev. Lett.* **80** 2586

[11] Butola A *et al* 2020 High space-bandwidth in quantitative phase imaging using partially spatially coherent digital holographic microscopy and a deep neural network *Opt. Express* **28** 36229–44

[12] Pandey N, Ghosh A and Khare K 2016 Two-dimensional phase unwrapping using the transport of intensity equation *Appl. Opt.* **55** 2418–25

[13] Wolf E and Marchand E 1964 Comparison of the Kirchhoff and the Rayleigh–Sommerfeld theories of diffraction at an aperture *JOSA* **54** 587–94

[14] Arfken G B and Weber H J 1999 *Mathematical Methods For Physicists* (College Park, MD: American Association of Physics Teachers)

[15] Hariharan P 2000 Phase-shifting interferometry: minimization of systematic errors *OptEn* **39** 967–69

[16] Hariharan P, Oreb B and Eiju T 1987 Digital phase-shifting interferometry: a simple error-compensating phase calculation algorithm *Appl. Opt.* **26** 2504–6

[17] Bhatt S, Butola A, Kanade S R, Kumar A and Mehta D S 2021 High-resolution single-shot phase-shifting interference microscopy using deep neural network for quantitative phase imaging of biological samples *J. Biophoton.* **14** e202000473

[18] Bhatt S, Butola A, Kanade S R, Kumar A and Mehta D S 2021 Single-shot phase-shifting interferometry using deep learning *European Conference on Biomedical Optics* (Washington, DC: Optical Society of America), ETu2A51

[19] Ahmad A *et al* 2020 Sub-nanometer height sensitivity by phase shifting interference microscopy under environmental fluctuations *Opt. Express* **28** 9340–58

[20] Rodriguez-Zurita G, Toto-Arellano N-I, Meneses-Fabian C and Vázquez-Castillo J F 2008 One-shot phase-shifting interferometry: five, seven, and nine interferograms *Opt. Lett.* **33** 2788–90

[21] Abdulhalim I 2012 Spatial and temporal coherence effects in interference microscopy and full-field optical coherence tomography *Ann. Phys.* **524** 787–804

[22] Ryabukho V, Lyakin D, Grebenyuk A and Klykov S 2013 Wiener–Khintchin theorem for spatial coherence of optical wave field *J. Opt.* **15** 025405

[23] Butola A *et al* 2020 High spatially sensitive quantitative phase imaging assisted with deep neural network for classification of human spermatozoa under stressed condition *Sci. Rep.* **10** 1–12

[24] Butola A *et al* 2021 Multimodal on-chip nanoscopy and quantitative phase imaging reveals the nanoscale morphology of liver sinusoidal endothelial cells *Proc. Natl Acad. Sci.* **118** e2115323118

[25] Kim M K 2011 Digital holographic microscopy' in digital holographic *Microscopy.* (Berlin: Springer), pp 149–90

[26] Kemper B, Vollmer A, von Bally G, Rommel C E and Schnekenburger J 2011 Simplified approach for quantitative digital holographic phase contrast imaging of living cells *J. Biomed. Opt.* **16** 026014

[27] Native A and Shaked N T 2017 Compact interferometric module for full-field interferometric phase microscopy with low spatial coherence illumination *Opt. Lett.* **42** 1492–94

[28] Choi Y, Yang T D, Lee K J and Choi W 2011 Full-field and single-shot quantitative phase microscopy using dynamic speckle illumination *Opt. Lett.* **36** 2465–68

[29] Popescu G 2011 *Quantitative Phase Imaging of Cells and Tissues* (New York: McGraw-Hill Education)

[30] Park Y, Depeursinge C and Popescu G 2018 Quantitative phase imaging in biomedicine *Nat. Photonics* **12** 578–89

[31] Dubey V *et al* 2019 Low coherence quantitative phase microscopy with machine learning model and Raman spectroscopy for the study of breast cancer cells and their classification *Appl. Opt.* **58** A112–A119

IOP Publishing

Quantitative Phase Microscopy and Tomography
Techniques using partially spatially coherent monochromatic light
Dalip Singh Mehta, Ankit Butola and Veena Singh

Chapter 3

Partially spatially coherent common-path quantitative phase microscopy

The current chapter describes the common-path quantitative phase microscopy (QPM) using partially spatially coherent light source in a detailed manner. Being less sensitive to environmental fluctuations and vibrations, the common-path techniques offer higher temporal phase stability. The chapter enlists the various techniques such as Gabor (in-line) holography, diffraction phase microscopy, Fourier phase microscopy, lateral shearing interferometric phase microscopy and Fresnel bi-prism based interferometric phase microscopy which provides a quantitative phase of the sample with higher temporal phase stability and a higher spatial phase sensitivity. Mostly these techniques are used in single-shot mode making them highly suited for time varying experiments.

3.1 Gabor (in-line) holography

In-line (common-path) holography was first developed by Dennis Gabor in the year 1948 [1], for which he accepted the Nobel Prize in the domain of Physics in the year 1971. Dennis Gabor was an electrical engineer, and he visualized the process of imaging as the method of transmission of information [1]. He published a paper in *Nature*, in 1948, titled 'A new microscopic principle' which utilized optical phase data in order to correct the abrasions in electron micrographs. Gabor explained the complete information about amplitudes and phases of light field [2]. He described the phenomenon of holography, where an object is illuminated with a beam of electrons which is first focused and thereafter diverges at a half angle *a*. Figure 3.1 shows the schematic ray diagram of Gabor's in-line holographic experiment. The object being placed at a small distance (d) in the path of incoming beam and a photographic plate is placed afterwards at a larger distance (probably an integer multiple of the d). The photographic film records the interference pattern produced due to superposition of the primary incoming wave front and the secondary wave front created by the object [2].

doi:10.1088/978-0-7503-3987-2ch3

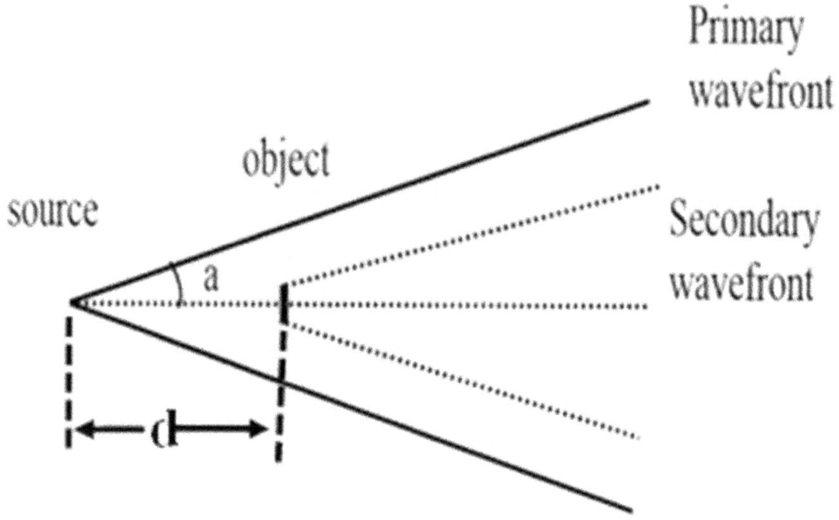

Figure 3.1. Illumination process described by Gabor.

Now, during the retrieval process if the developed/printed photograph also known as a hologram is illuminated with the optical imitation of the incoming beam then the user expects to see the original object as if it is kept in place. Conventional holography in general involves two steps; the first step includes recording the hologram of the object on a photographic plate and the second step includes recovering the object from the developed hologram by illuminating it with the reference wave. Early versions of holographic systems were realized using incoherent light sources, such as spectrally filtered light from a tungsten–halogen lamp, hence the holograms were free from spurious fringes and speckles owing to the low temporal coherence nature of the light source. But using low coherent light it was quite difficult to realize holographic set-ups, particularly off-axis systems like Mach–Zehnder system. Since the development of lasers in the 1960s most of the holographic experiments (in-line or off-axis) have been realized using highly coherent light (spatially and temporally) like lasers, hence it is very easy to realize interferometric and holographic experimental set-ups of any optical configuration, i.e., on-axis or off-axis. Because of lasers' high spatio-temporal coherence, the recorded holograms/interferograms are cluttered with spurious fringes and speckle noise, leading to poor reconstruction of images and phase maps. To avoid these problems a partially spatially coherent laser (PSCL) light can be used. The Gabor's in-line holographic systems can be realized using PSCL as shown in figures 3.2(a) and (b) to achieve a speckle-free quantitative phase image of the object. Figure 3.2(a) shows the Gabor's (in-line) holographic experimental set-up using PSCL light with spherical wave front, and figure 3.3(b) shows using a collimated beam, i.e., plane wave front. First, the laser light passes through an RGG (rotating ground glass) and the temporally fluctuating scattered laser light is collected by the microscope objective (MO) lens which focusses light at the pinhole (PH). Light emerging

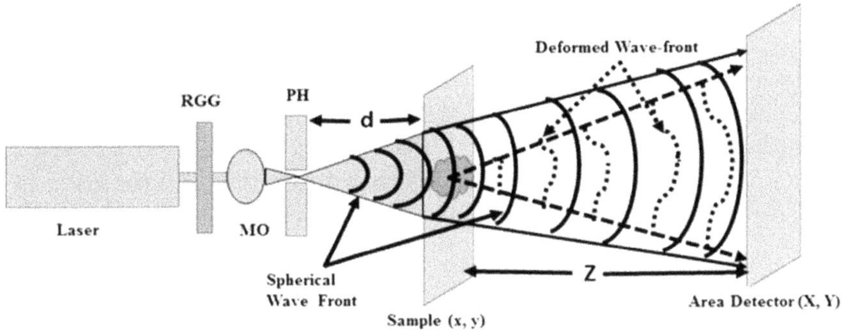

(a)

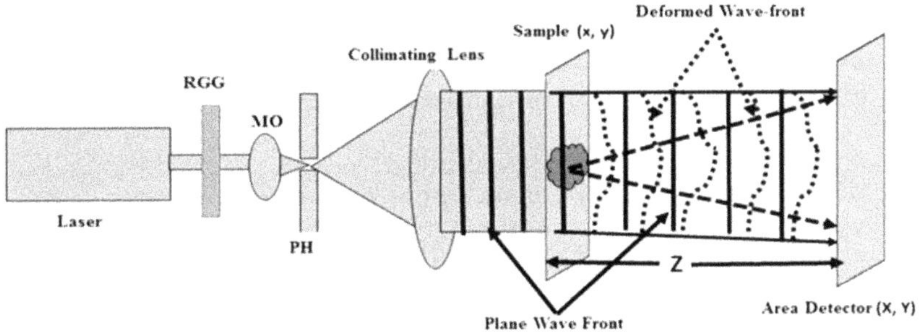

(b)

Figure 3.2. Illustrating the Gabor (in-line) holographic experimental set-up using PSCL light; RGG—rotating ground glass; MO—microscope objective; PH—pinhole; (a) spherical wave front and (b) plane wave front.

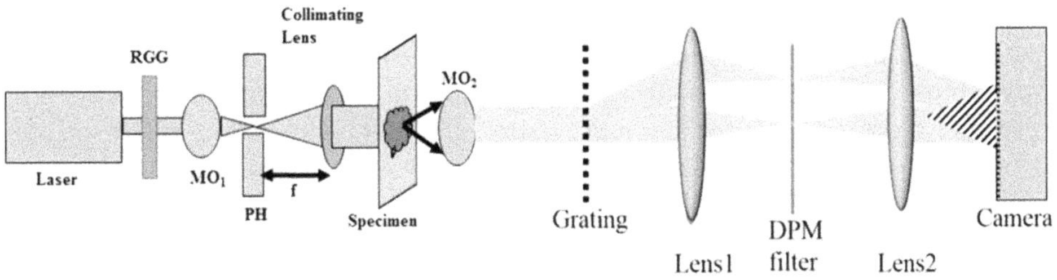

Figure 3.3. Schematic diagram illustrating the optical configuration for diffraction phase microscopy system with partially spatially coherent light.

from the PH will be a diverging wave front (figure 3.2(a)) interrogating with the sample. In figure 3.2(b) by means of placing a collimating lens exactly at focal length of the lens from PH, a collimated plane wave front can be generated to interact with the object. The combination of RGG, MO and PH can make the coherent light into a PSCL and by means of controlling the size of pinhole the spatial coherence of the light can be tuned. The partially coherent diverging spherical wave front illuminates the object placed onto the glass substrate. The spherical wave front is deformed after transmitting in the areas of object while a part of it remains undisturbed without a sample. Both the wave fronts propagate together in the common path and produce interference or a hologram which is recorded by an area detector like a charged coupled device (CCD) and accumulated in the computer. The final images are computed using a suitable program.

In figure 3.2 the electromagnetic field interrogating with a sample given by $S(x,y)$ and the reference fields given by $R(x,y)$ can be expressed, respectively as [3];

$$S(x, y) = s(x, y)\exp\left[-j\phi_s(x, y)\right] \tag{3.1}$$

$$R(x, y) = r(x, y)\exp\left[-j\phi_r(x, y)\right] \tag{3.2}$$

where $\phi_s(x, y)$ and $\phi_r(x, y)$ are the phase of the sample wave front and the reference wave front, respectively. As shown in figures 3.2(a) and (b), in the region of superposition the interference of sample and reference fields takes place and the resulting intensity distribution of hologram recorded onto the camera CCD can be stated as:

$$\begin{aligned} I(x, y) &= \lfloor S(x, y) + R(x, y)\rfloor^2 \\ &= \lfloor S(x, y)\rfloor^2 + \lfloor R(x, y)\rfloor^2 + 2S(x, y)R(x, y) \\ &\quad \cos\left[\phi_s(x, y) - \phi_r(x, y)\right] \end{aligned} \tag{3.3}$$

Using equation (3.3) by means of using a suitable reconstruction method, such as, Fresnel transform, Fourier transform or phase shifting algorithm, the phase map of the light field interrogated with the sample can be reconstructed. Holography finds its application in optical metrology as well as in the field of biophotonics, as it helps in measurement of phase information with high accuracy. In the field of biophotonics, quantitative phase interferometry has been utilized for the determination of phase with nanoscale accuracy since the 1950s. Phase information is related to various biological parameters, thereby helping in the computation of various parameters such as RI (refractive index), dry cell mass etc [4, 5]. In the past decade, interferometry has been frequently utilized for providing quantitative parameters of the specimen. Quantitative phase microscopy has been employed in various applications in the field of biophotonics. Interferometry is usually implemented using off-axis and in-line configurations. An off-axis system provides separate paths for the incoming reference beam and object beam, i.e., the reference and object beams pass through different paths and sets of different optical components which makes them more sensitive to the environment fluctuations and vibrations. An in-line configuration system provides the same path to the reference and object beams, i.e., both the beams go through an identical path and through the identical set of optical components. In-line systems are less sensitive to environment noise, fluctuations and vibrations.

The experimental set-ups shown in figures 3.2(a) and (b) are lensless holographic systems, therefore, the system has limited resolution and very poor magnification. For biological applications where the sizes of the object features are in the range of microns, then a highly magnifying lens like an MO is required to gather the scattered field from the object, and a magnified image with interference pattern is recorded onto the chip of the CCD camera. One of the most important drawbacks which affects the accuracy of the quantitative phase microscopic systems is their sensitivity to environmental noise and fluctuations. In addition, if the holographic microscope set-up is also realized in common-path, on-axis then the entire system becomes free from surrounding vibrations, along with high resolution and high magnification. The best possible way to solve this problem is to convert a conventional optical microscope into a common-path quantitative phase microscope. In the following section we describe two such systems.

3.2 Diffraction phase microscopy

Diffraction phase microscopy (DPM) is a sort of common-path QPM system which helps in getting rid of the environment noise and fluctuations [6, 7]. The DPM is realized using a conventional microscope to which a common-path interferometric system is attached, while utilizing the entire microscopic system for imaging and illumination. The DPM is implemented in common-path configuration which helps in cancelling out most of the vibrational noises. Further, the system is single-shot which means it records the interferogram using a single image capture. The method itself is non-destructive, non-invasive and non-contact and thus can be accomplished in numerous types of applications as well as the dynamic applications. The system has the capability of operating in both reflection as well as transmission mode such that it can be easily employed for the visualization of transparent as well as opaque specimens. Thus, there are many uses of DPM which include measurement of the dynamics of the biological specimens, wet etching of semiconductor and photochemical etching processes, and water droplets evaporation, and defect detection in semiconductor wafer etc.

Figure 3.3 shows a typical DPM system using light which is partially spatially coherent. The light from the laser passes through RGG and scattered light is focussed onto the pinhole using MO_1. A collimating lens further collimates the light field emerging from the PH which illuminates the specimen. The scattered field from the sample is collected by high numerical aperture MO_2. The image will be further relayed by a tube lens. The DPM system utilizes a diffraction grating in combination with a symmetric/asymmetric 4f lens system as shown in figure 3.3. The DPM system can be positioned at the end interface of any conventional microscopic system. The light from the conventional microscopic system falls on the grating which diffracts the incoming light into multiple orders, i.e., 0th, ±1st etc. Out of these various diffracted orders 0th and +1st orders are employed to generate the interference pattern with the assistance of a 4f lens configuration. Since all the orders are generated from the same beam, therefore, containing both low and high spatial frequency components including object information. In the DPM system, the generated 0th order from the incoming light beam acts as the sample/object beam and the generated 1st order is filtered to produce the reference beam. The reference beam is filtered by using DPM filter, which can be visualized as a set of small PHs as well as large aperture.

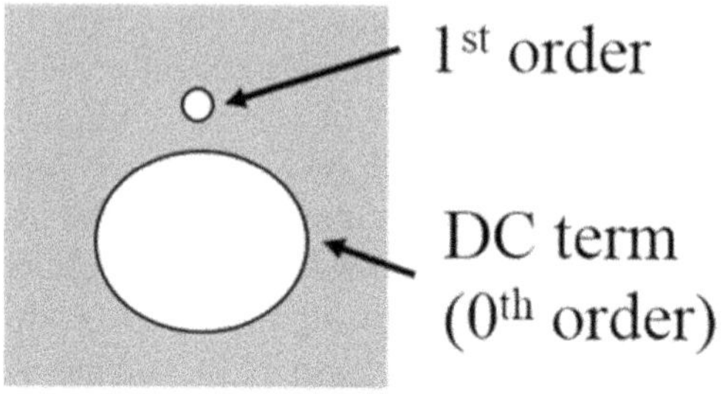

Figure 3.4. Schematic diagram of the DPM filter.

Figure 3.4 shows the DPM filter in which the 0th order passes through it unaltered but 1st order is cleaned utilizing a PH in order to pass only the DC term which serves as reference beam. Figure 3.3 demonstrates the diagram of the DPM system. It comprises a grating and a 4f lens system using lens1 and lens2. Lens1 performs the Fourier transform of the information and the DPM filter is placed in the Fourier plane. The lens2 of the 4f configuration again takes the Fourier transform of the incoming beam and the two beams thus generated are interfered at the image plane (IP). The interferogram is captured with the aid of a CCD camera. The interferogram is processed via Fourier transform to reconstruct the phase map of the object. Using this phase map other biophysical parameters can be computed. The spatial resolution and magnification of the aforementioned DPM system is very high and as good as a conventional microscope.

3.3 Fourier phase microscopy

Another version of high-resolution common-path QPM is Fourier phase microscopic (FPM) system, which works on the principle of breakdown of the data into two components spatially such that these two components can be easily controlled with help of SLM (spatial light modulator) and can be phase shifted relating to one another. It can be described as a type of phase shifting common-path quantitative phase imaging technique. It unites the advantages of phase shifting technique along with higher stability of common-path systems. It helps to transform a conventional qualitative microscope into a quantitative phase microscope. The system offers high stability in addition to high transverse resolution. The technique consists of attaching the system at the end interface of a conventional microscopic system. The schematic diagram for FPM can be visualized in figure 3.5. They have utilized inverted microscope (Axiovert 35, Carl Zeiss, Inc.) with super-luminescent diode illumination light source (809 ± 20 nm) which was coupled into a single mode fibre and further collimated in a way that the plane wave front illuminates the object. The amplified image formed at the IP appears to be formed by virtual point source (VPS). The correcting lens (CL) is placed at the IP with focal length in a way that it pictures VPS at infinity. A Fourier lens denoted by FL is used to produce the Fourier transform of the image field on the PPM (programmable phase modulator, i.e. a 2D liquid crystal array). The polarizer is used to produce field polarization parallel to the axis of the liquid crystal (LC). As in figure 3.5, the PPM has detailed

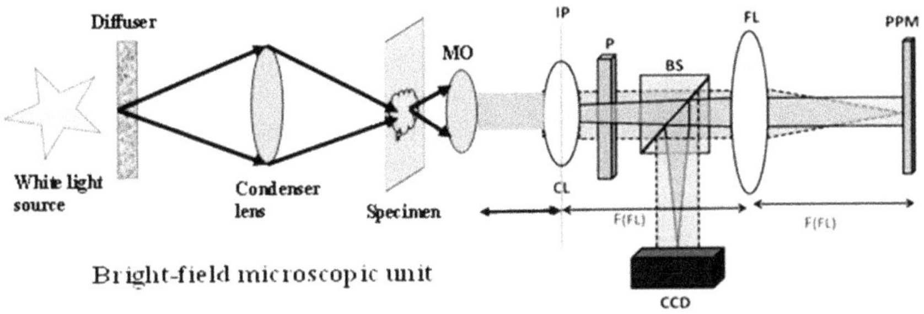

Figure 3.5. Experimental set-up of Fourier phase microscopy (re-drawn from [8]).

control over the phase of light which is reflected by its surface and is used to produce four phase shifts in increments of $\pi/2$ in accordance with average field, the same as in conventional phase shifting interferometry (PSI) technique.

FPM usually employs lasers as a source of illumination, but the use of lasers leads to degradation of results as well as results in poor resolution of the system. In reference [9] the authors have employed white light source for providing uniform illumination to the sample and minimizing the speckles by the use of broadband light source.

3.4 Lateral shearing interferometric phase microscopy

The lateral shearing interferometer was first introduced by W J Bates in the year 1946 where he described the sphericity of the optical wave front using shear or lateral displacement [10]. Lateral shearing interferometers have been implemented by various methods involving the use of double grating [11], wedge and square prisms [12, 13], parallel plates, wedge plates [14–17] etc. Later, Murty [15] generated the lateral shearing interferometer using a slightly wedged plate or a parallel glass plate. He utilised a He–Ne laser along with the parallel glass plate to generate interference fringes due to the superposition of beams reflected from the front surface and back surface of the glass plate. Lateral shearing interferometers have also been used in the field of precision measurement such as surface profile measurement systems. Authors in reference [18] have utilized a single birefringent crystal for the generation of shear amount. They realized good contrast fringes by means of using circularly polarized light giving the system a common path with good stability and accuracy. Since, then the lateral shearing interferometer has found its applications in various fields such as wave front sensing, free form optics and in phase microscopy. Recently, lateral shearing interferometric phase microscopy has been used in the field of biophotonics and biomedicine in phase measurement of biological specimens [19–24] etc. Lateral shearing interferometric systems have an advantage over off-axis systems as they are common-path in nature. The beams, both reference and object, travel through the same collection of optical elements, making them less vulnerable to environmental vibrations and fluctuations contrasted with their off-axis counterparts such as Michelson and Mach–Zehnder interferometers. Off-axis systems have the disadvantage of higher sensitivity to phase noise as the quantitative phase information is subjected to air fluctuations and mechanical vibrations. The common-path nature of the lateral shearing interferometers makes them highly temporally stable and compact. There have been many modifications to lateral shearing interferometers to be used as QPM for biological applications. Ferraro *et al* [19] combined the technique of lateral shearing interferometer with digital holographic microscope (DHM) to measure the quantitative phase information. He even demonstrated the use of the technique on biological specimens such as a mouse cell line.

The techniques and configurations discussed above employ lasers as the illumination source. Lasers are highly coherent in nature and have the characteristic of high spatial coherence, high temporal coherence and long coherence length. However, this high coherence nature of lasers leads to the disadvantage of generating coherent artefacts such as speckles. Speckles are generated when imaging is carried out in a scattering environment. Speckles are an unwanted pattern which is generated in the scattering environment due to the phase delay introduced in mutually coherent photons whose interference is at the detection plane. In addition to the speckles, the high spatial-temporal coherence of lasers leads to unwanted spurious fringes. These spurious or unwanted interference fringes are mainly the result of multiple reflections arising from the multilayers of the optical components employed in the set-up. The interference of these multiple reflections leads to the presence of spurious fringes in the results. These speckle artefacts and spurious/false fringes behave as noise in the results and decrease the accuracy of the entire system. When lasers are employed in applications such as quantitative phase measurement, these phase noise artefacts lead to overall decrease in accuracy of the system as well loss of information. To overcome this disadvantage of lasers, partially spatially coherent sources are used in biological applications of quantitative phase imaging. Partially spatially coherent sources have low spatial coherence which leads to a decrease in the spurious fringes and speckles. In reference [20] the authors employed a light source which is partially spatially coherent in combination with lateral shearing interferometer to generate uniform illumination and less spurious interference fringes. The authors used a rotating diffuser placed in front of a He–Ne laser which provided speckle-free imaging and uniform illumination of the object/sample. The experimental set-up used by the authors is shown in figure 3.6.

The experimental set-up in figure 3.6 shows the generation of light source which is partially spatially coherent using an RGG diffuser. The light gets scattered on using a moving diffuser and the scattered light is collected with the assistance of a microscopic objective MO_1 which focusses light onto the PH. Scattered light emanating from the PH is further collimated using a lens. The specimen is placed in front of the second MO lens MO_2 which collects the scattered light, and an image of the object is further relayed by a tube lens. A 4 mm shear plate is used to generate the interference pattern which is recorded on CCD_1. The authors in reference [20] have employed two CCD cameras to record interference pattern and bright field image of the sample simultaneously. The CCD_2 in the above image is employed to capture the bright field image. The synthesized source is coupled to

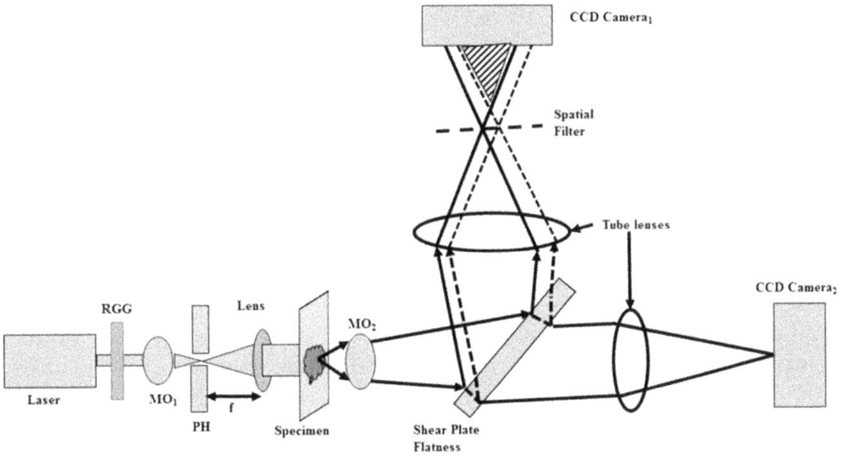

Figure 3.6. Schematic diagram illustrating the optical configuration for a phase microscopic system with lateral shearing interferometer and partially spatially coherent light.

the lateral shearing interferometer. The interference pattern is generated due to the superposition of wave fronts which gets reflected from the front surface and back surface of the shear plate. The optical path difference (OPD) between the two interfering wave fronts depends on the shear plate thickness h, the refractive index denoted by n and the incident angle denoted by Θ, and is expressed by the following formula; $OPD = \frac{2n\,h}{\sqrt{n^2 - \sin^2\theta}}$. To obtain high contrast interference fringes the OPD between the two beams must be less than or equal to the temporal coherence length l_c of the light source, OPD $\leqslant l_c = 0.44\lambda^2/\Delta\lambda$, where λ is the peak wavelength and $\Delta\lambda$ is the spectral full-width half maximum of the light. In the case of shear plate interferometers, the OPD between the two interfering beams is in the range of several mm to cms. Therefore, only highly monochromatic light sources such as lasers can be used and light sources like LEDs or white light cannot be used. Further, the flatness of both the surfaces of shear plate must be $\geqslant \lambda/10$ and all the other optical components used in the set-up shown above must be of high quality, and free from all kind of aberrations. If these conditions of the interference are satisfied then the difference in the phase values between the two interfering beams is given by the following expression;

$$\varnothing(x, y) = \frac{2\pi}{\lambda}OPD = \frac{4\pi nh}{\lambda\sqrt{n^2 - \sin^2\theta}} \tag{3.4}$$

The interference of the two beams is given by the equation

$$I(x, y) = I_a(x, y) + I_b(x, y) + 2\sqrt{I_a(x, y)I_b(x, y)}\,\gamma_{12}(\tau)\cos\Delta\Phi(x, y) \tag{3.5}$$

In the above equation I_a and I_b are the intensities for the two beams (reference and object), $\gamma_{12}(\tau)$ is the coherence envelope, $\Delta\Phi$ denotes phase difference. The phase difference is given by the equation below, where n_1 and n_2 are the refractive indexes of the cell and medium.

$$\Phi(x, y) = \frac{2\pi}{\lambda}(n_1(x, y) - n_2(x, y))*h(x, y) \tag{3.6}$$

In the above equation, $h(x, y)$ stands for the height of the sample. Since in the current set-up both the reference and object beam travel through the sample, the system faces the twin image problem. This can be overcome with the use of sparse samples (figure 3.7).

The authors in [20] imaged red blood cells and reconstructed their phase image. They demonstrated the use of partially spatially coherent source and determined spatial phase sensitivity. The authors reported the spatial phase sensitivity 0.7567 ± 0.3625 rad and 0.1876 ± 0.0938 rad for laser and partially spatially coherent source, respectively (figure 3.8).

The common-path nature of the lateral shearing interferometer leads to higher stability. The common-path system is less vulnerable to environmental fluctuations and results in stable interferograms. The authors measured the temporal stability of their developed system by capturing the sequence of interferograms at a frame rate of 15 frames/s. The phase difference was determined with respect to the first frame and standard deviation of the calculated phase difference was determined for 10^4 random pixels and obtained mean standard deviation as 6.6 mrad (figure 3.9).

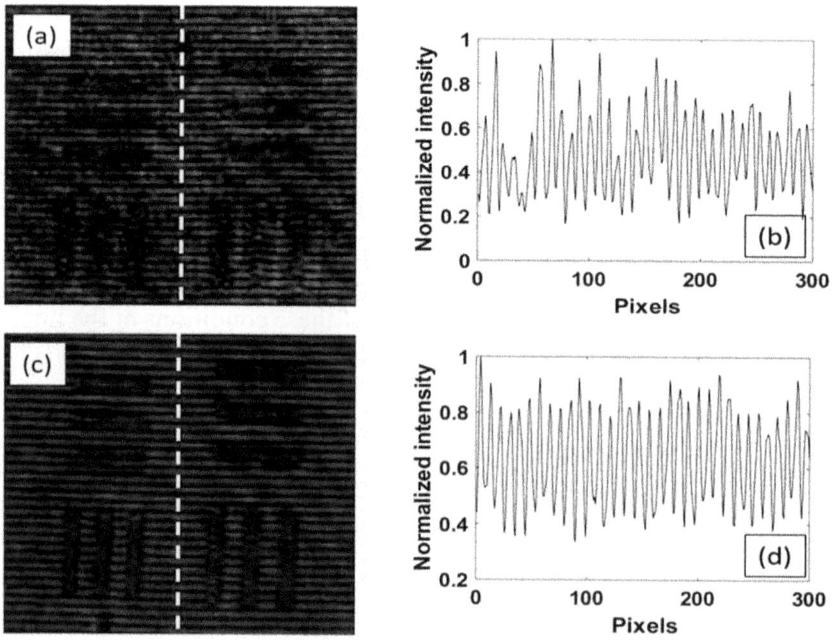

Figure 3.7. Interferogram using shear interferometric set-up; (a) with coherent light, (b) corresponding line profile, and (c) with partially spatially coherent light and (d) line profile.

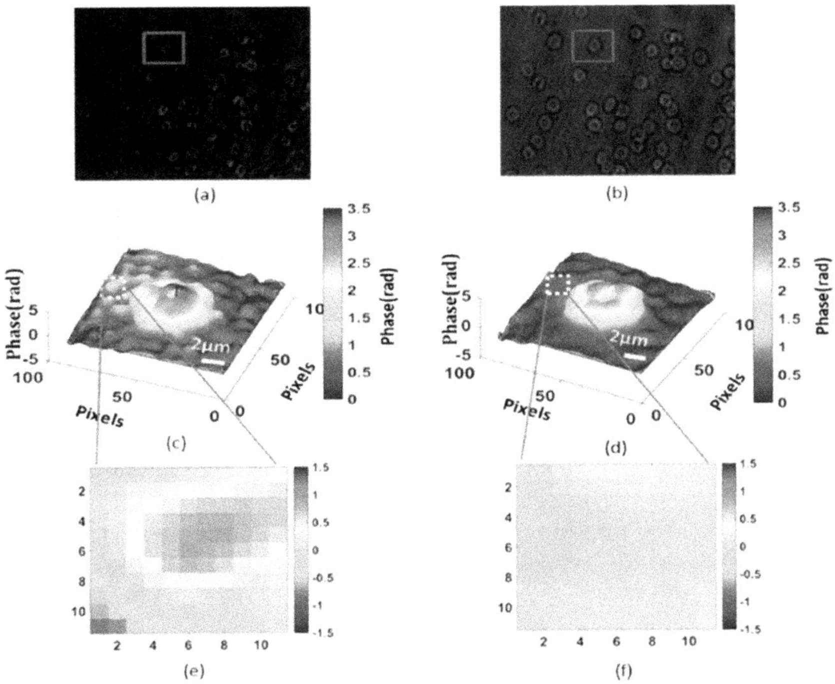

Figure 3.8. (a) Captured interferogram for laser; (b) captured interferogram for partially spatially coherent light source; (c) and (d) phase profile reconstructed for red blood cells (RMCs) using source as laser and partially spatial light source; (e) and (f) spatial phase sensitivity of laser and partially spatial light source. Reproduced from [20], copyright (2019) with permission from Elsevier.

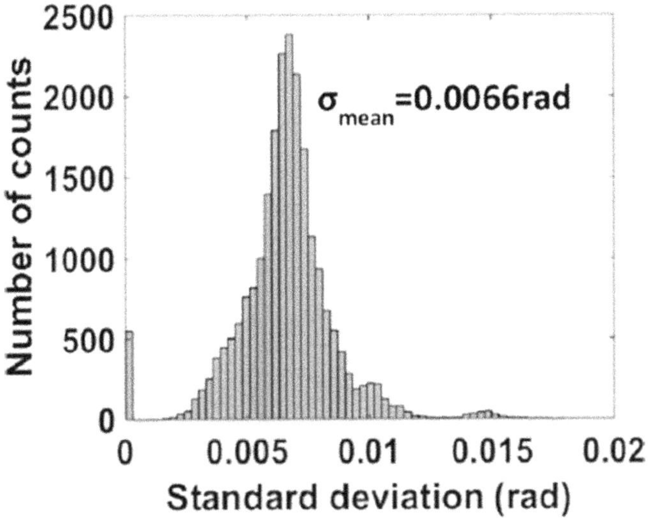

Figure 3.9. Histogram for measuring the temporal stability of the system. Reproduced from [20], copyright (2019) with permission from Elsevier.

3.5 Fresnel bi-prism based interferometric phase microscopy

As discussed in the previous section, common-path interferometers are highly stable compared to their off-axis counterparts such as Mach–Zehnder and Michelson interferometers. They have high temporal stability as well as being more robust for time dependent studies. In common-path systems, as in lateral shearing interferometer, the reference and object beam travel through the same set of optical components. The system faces the disadvantage of twin images. This disadvantage of twin images can be surmounted with the help of the production of the reference beam using spatial filtering method. The authors in reference [25] have employed a τ interferometer as a common-path system and used a PH in the Fourier plane to generate the reference beam. As in DPM, grating has been employed to generate two beams and one of the beams is spatially filtered to generate the reference beam [26]. But the problem is that the grating generates multiple orders but only two orders are used, leading to a large amount of power loss in the other orders.

In addition to spatial filtering method, self-referencing technique is also employed for the reference beam. In self-referencing technique the reference beam is produced by placing the sample only in half of the beam while the remaining half acts as reference beam. In references [27, 28] the authors used a single-beam splitter to split up and merge the wave front and they placed the sample only in half of the beam. Even if lateral shearing method is employed in combination with spatial filtering [29–31] to solve the twin image problem, the glass plate only reflects 4% of the power, thus bringing about a large amount of loss of power.

To overcome these disadvantages of other common-path systems Fresnel bi-prism is employed. Fresnel bi-prism divides the travelling beam equally into two parts [32–34]. The generated beams interfere to give the linear interference fringe pattern with fringe spacing given by; $d = \frac{L\lambda}{b(n-1)\alpha}$, where L and b describe the distances between the virtual sources to the sensor plane and the bi-prism, respectively, n is the refractive index and α is angle of bi-prism, as shown in figure 3.10. The system overcomes the twin image problem using both the methods described above, i.e., spatial filtering and self-referencing. TheFresnel bi-prism provides the easiest means of creating two beams of equal intensity and thus provides good fringe contrast which is a requirement for interferometry. The system uses an asymmetric 4f imaging configuration to attain

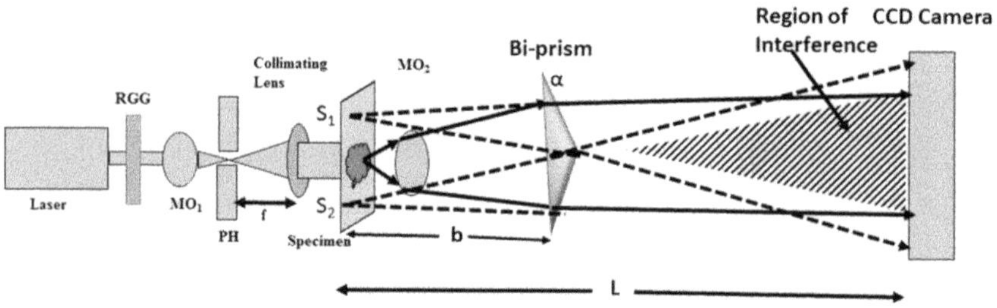

Figure 3.10. Schematic diagram illustrating the optical configuration for a phase microscopic system with partially spatially coherent light using Fresnel bi-prism.

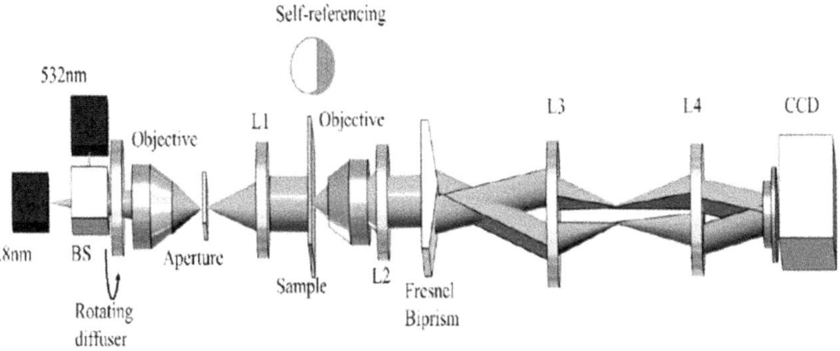

Figure 3.11. The schematic diagram of the experimental set-up using self-referencing mode. Copyright IOP Publishing. [32] Reproduced with permission. All rights reserved.

appropriate fringe density on the IP. The advantage of the Fresnel bi-prism set-up is that it is quite effortless to align and has least loss of power. Its common-path nature gives rise to higher temporal stability and makes it useful to image sparse as well as dense samples. In reference [32] the authors implemented the speckle-free Fresnel bi-prism interferometric system. They employed partially spatially coherent light source which helped them to achieve speckle-free image quality in addition to high temporal stability and less power loss. A Fresnel bi-prism was used to generate two beams and the system was used in self-referencing mode. The object/sample was positioned in such a way that it occupied half of the incoming beam, making the results free of the overlapping image problem. The experimental set-up for a partially spatially coherent Fresnel bi-prism interferometer is shown in figure 3.11. The authors have demonstrated the interferometer using two wavelengths 532 nm and 632.8 nm. The lasers were merged using a beam splitter and a rotating diffuser was used to create the partially spatially coherent source. The rotating diffuser scatters the incoming light which is collected using a 10× MO, which was further collimated using a lens L1 passed through the sample and was gathered by a 20× MO. The Fresnel bi-prism splits up the incoming beam into halves, i.e. reference beam and object beam. The generated beams travel at an angle to each other which interfere each other on the camera plane. An asymmetric 4f system is utilized to interfere the beam on the IP as it provides an extra magnification to realize appropriate fringe density. A color CCD camera is used to capture the images. The sample is placed in a way that it occupies only half of the beam so half the sample acts as object beam while the remaining half acts as reference beam [32].

3.6 Spatial phase sensitivity

To evaluate spatial phase sensitivity of a Fresnel bi-prism interferometric system using a partially spatially coherent light source the interferograms were recorded. Figures 3.12(a) and (b) display the interferograms recorded for laser and partially spatially coherent light source. A coherent laser source provides non-uniform fringe contrast due to non-uniform illumination, while a partially spatially coherent source provides uniform fringe contrast throughout the IP due to speckle-free and uniform illumination. Figure 3.12(c) demonstrates the spatial phase sensitivity; the black

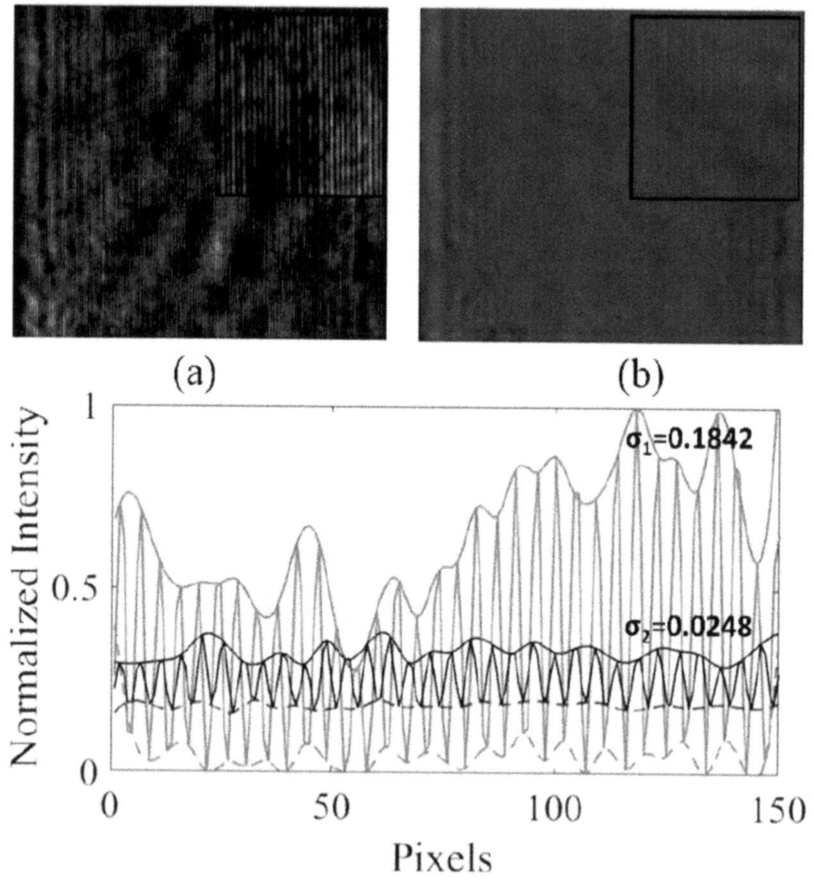

Figure 3.12. (a) and (b) recorded interferogram for laser and partially spatially coherent source, respectively. (c) Fringe intensity variation with respect to spatial location. Copyright IOP Publishing. [32] Reproduced with permission. All rights reserved.

curve is for partially spatially coherent source and the red curve for coherent laser. The figure shows the fringe intensity values plotted with respect to the spatial location. The intensity minimas do not vary much compared to intensity maximas values. Intensity standard deviation is determined to be 0.1842 for laser and 0.02 for partially spatially coherent light source, which is an order of magnitude better than that of laser. This variation in the fringe contrast results in spatial phase noise.

3.7 Temporal phase stability

Temporal stability of interferometric depends on how sensitive it is to the environmental fluctuations and vibrations. The common-path systems tend to have higher temporal stability compared to off-axis systems. A partially spatially coherent Fresnel bi-prism interferometric phase microscopic system has high temporal stability. The temporal stability of the system was measured by recording a sequence of interferograms at 15 frames/s over a period of 10 s. Thereafter, phase difference was determined according to the first frame of the sequence and then standard deviation of the phase difference was evaluated at 4×10^4 pixel points randomly. Their mean was noted for the given set-up. Figure 3.13 displays the histogram plotted for temporal stability. The mean value for laser was found to be 7 mrad and for partially spatial source was found to be 7.8 mrad. Their comparable values shows that the partially spatially coherent system helps us in achieving comparable

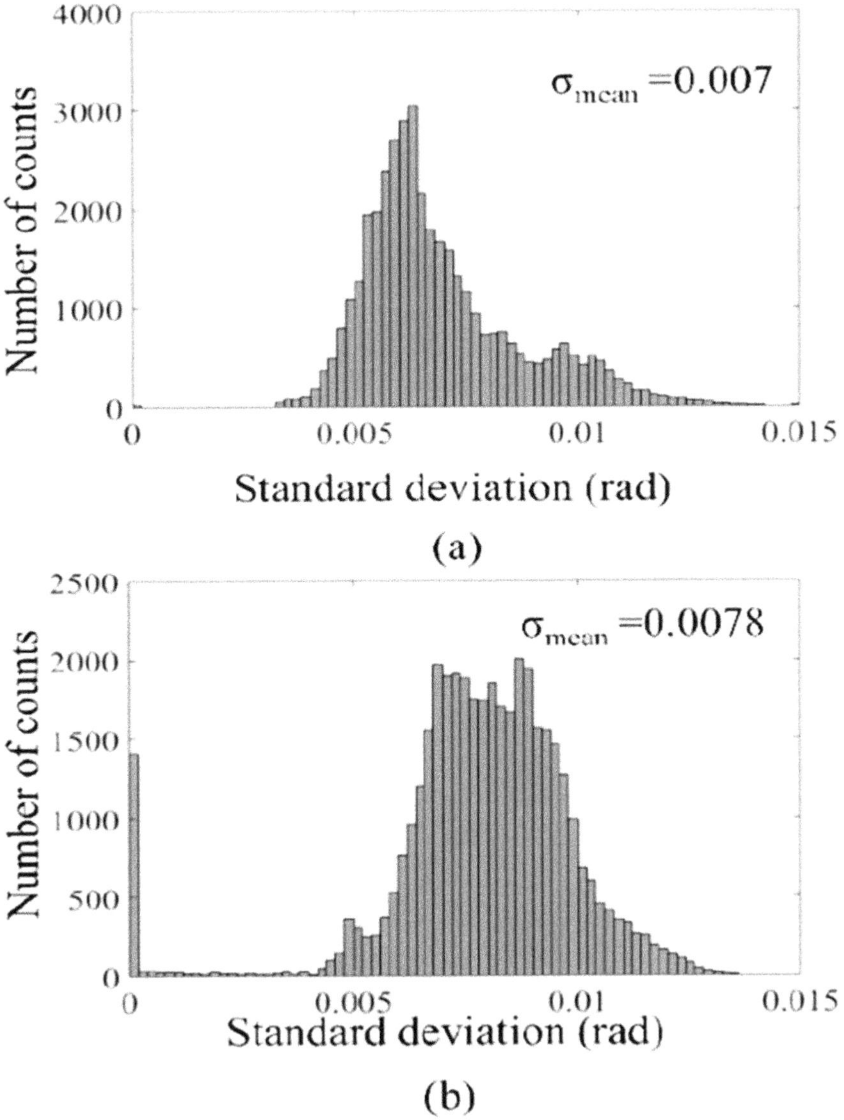

Figure 3.13. Histogram for temporal phase stability (a) laser (b) partially spatially coherent light source.

temporal stability with uniform illumination, high spatial stability and reduced speckles.

3.8 Resolution of the system

The variation in the fringe contrast, illumination non-uniformity, and spurious or false fringes due to coherent laser gives rise to spatial phase noise; this acts as background noise in the reconstructed phase [35]. This further leads to inaccuracy in the results, loss of information and decrease in resolution in the recorded images. This can easily be visualized in figure 3.14 which displays the interferograms captured for laser and partially spatially coherent light source using RBC. The illumination non-uniformity and spurious fringes in interferogram in laser source lead to the loss of phase information in figures 3.14(e) and (d).

The effects of fully spatial and partially spatially coherent light in diffraction phase tomography [3, 36] and non-interferometric phase microscopy [37] with partial spatial coherence has been demonstrated. In all these techniques it has been found that speckle-free, cross-talk free phase imaging of biological samples has been observed with higher accuracy in phase measurement. An innovative scheme for generating partially spatially coherent illumination for quantitative phase microscopy was also demonstrated [38, 39]. It has been found that partially spatially coherent light has several advantages, e.g., speckle-free imaging, removing halo effects and improved optical sectioning. Various methods and strategies for significantly reducing speckle noise in digital holography has been described in a review article [40].

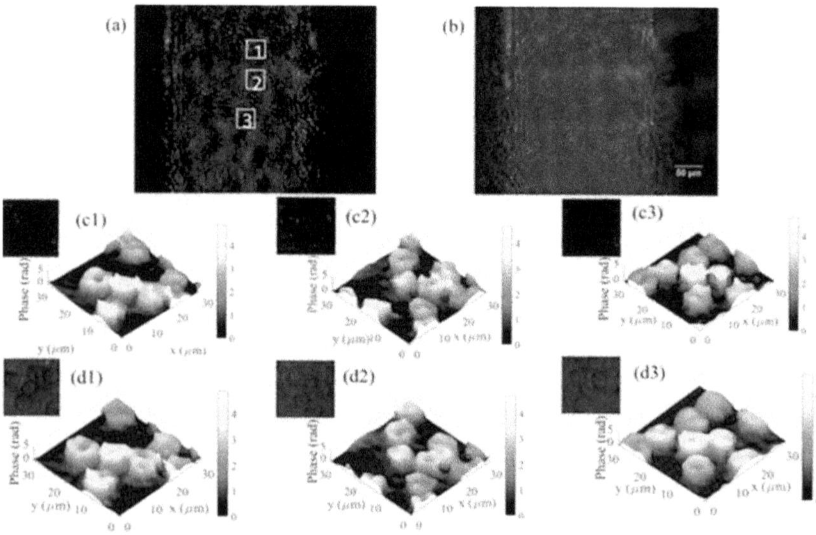

Figure 3.14. (a) and (b) captured interferogram using coherent laser source and partially spatially coherent light source, respectively, (c) and (d) reconstructed phase map of laser and partially spatially coherent source. Copyright IOP Publishing. [32] Reproduced with permission. All rights reserved.

In summary, various common-path QPM configurations are described, and an order of magnitude temporal phase stability is obtained compared to their off-axis counterparts. In addition, if the partially spatially coherent light is used in common-path QPM systems then spatial phase sensitivity can also be enhanced considerably. By means of utilizing both of these concepts the common-path QPM systems can be used for phase microscopy of dynamic biological cells.

References

[1] Gabor D 1946 Theory of communication *J. Inst. Electr. Eng.* **93** 429–57

[2] Gabor D 1948 A new microscopic principle *Nature* **161** 777–78

[3] Gabor D, Stroke G W, Brumm D, Funkhouser A and Labeyrie A 1965 Reconstruction of phase objects by holography *Nature* **208** 1159–62

[4] Davies H G and Wilkins M H F 1952 Interference microscopy and mass determination *Nature* **169** 541

[5] Barer R 1952 Interference microscopy and mass determination *Nature* **169** 366–7

[6] Bhaduri B, Edwards C, Pham H, Zhou R, Nguyen T H, Goddard L L and Popescu G 2014 Diffraction phase microscopy: principles and applications in materials and life sciences *Adv. Opt. Photonics* **6** 57–119

[7] Talaikova N A, Popov A P, Kalyanov A L, Ryabukho V P and Meglinski I V 2017 Dual mode diffraction phase microscopy for quantitative functional assessment of biological cells *Laser Phys. Lett.* **14** 105601

[8] Popescu G, Deflores L P, Vaughan J C, Badizadegan K, Iwai H, Dasari R R and Feld M S 2004 Fourier phase microscopy for investigation of biological structures and dynamics *Opt. Lett.* **29** 2503–5

[9] Bhaduri B, Tangella K and Popescu G 2013 Fourier phase microscopy with white light *Biomed. Opt. Express* **4** 1434–41

[10] Bates W J 1947 A wavefront shearing interferometer *Proc. Phys. Soc. (1926–48)* **59** 940

[11] Betend-Bon J-P, Wosinski L and Breidne M 1992 Double grating phase stepping interferometry for testing aspherics *Pure Appl. Opt.: J. Eur. Opt. Soc. A* **1** 55–69

[12] Xu D and Chen S 1995 Novel wedge plate beam tester *Opt. Eng.* **34** 169–72

[13] Kim S W, Cho W J and Kim B C 1998 Lateral shearing interferometer using square prisms for optical testing of aspheric lenses *Meas. Sci. Technol.* **9** 1129–36.

[14] Mehta D S, Singh P, Faridi M S, Mirza S and Shakher C 2005 Two-wavelength lateral shearing interferometry *Opt. Eng.* **44** 085603

[15] Murty M V R K 1964 Use of a single-plane parallel plate as a lateral shearing interferometer with a visible gas laser source *Appl. Opt.* **3** 531–4

[16] Nomura T, Kamiya K, Okuda S, Miyashiro H, Yoshikawa K and Tashiro H 1998 Shape measurements of mirror surfaces with a lateral shearing interferometer during machine running *Precis. Eng.* **22** 185–9

[17] Mehta D S, Dubey S K, Hossain M M and Shakher C 2005 Simple multifrequency and phase-shifting fringe projection system based on two wavelength lateral shearing interferometry for three-dimensional profilometry *Appl. Opt.* **44** 7515–21

[18] Liu X, Gao Y and Chang M 2009 A new lateral shearing interferometer for precision surface measurement *Opt. Lasers Eng.* **47** 926–34

[19] Ferraro P, Alferi D, De Nicola S, De Petrocellis L, Finizio A and Pierattini G 2006 Quantitative phase-contrast microscopy by a lateral shear approach to digital holographic image reconstruction *Opt. Lett.* **31** 1405–7

[20] Tayal S, Usmani K, Singh V, Dubey V and Mehta D S 2019 Speckle-free quantitative phase and amplitude imaging using common-path lateral shearing interference microscope with pseudo-thermal light source illumination *Optik* **180** 991–96

[21] Ma Y, Li J, Zhang P, Li T, Xi J, Di and Zhao J 2017 Lateral shearing common-path digital holographic microscopy based on a slightly trapezoid Sagnac interferometer *Opt. Express* **25** 13659–67

[22] Bon P, Maucort G, Wattellier B and Monneret S 2009 Quadriwave lateral shearing interferometry for quantitative phase microscopy of living cells *Opt. Express* **17** 13080–94

[23] Zhang J, Dai S, Ma C, Xi T, Di J and Zhao J 2021 A review of common-path off-axis digital holography: towards high stable optical instrument manufacturing *Light: Adv. Manufact.* **2** 333–49

[24] Tayal S, Saxena A, Singh V, Kaur T, Singh N, Khare K and Mehta D S 2022 Design and development of integrated TIRF and common-path quantitative phase microscopic health care system with high stability *Opt. Lasers Eng.* **155** 107057

[25] Shaked N T 2012 Quantitative phase microscopy of biological samples using a portable interferometer *Opt. Lett.* **37** 2016–18

[26] Popescu G, Ikeda T, Dasari R R and Feld M S 2006 Diffraction phase microscopy for quantifying cell structure and dynamics *Opt. Lett.* **31** 6

[27] Qu W, Bhattacharya K, Choo C O, Yu Y and Asundi A 2009 Transmission digital holographic microscopy based on a beam-splitter cube interferometer *Appl. Opt.* **48** 2778–83

[28] Qu W, Chee O C, Yu Y and Asundi A 2011 Characterization and inspection of microlens array by single cube beam splitter microscopy *Appl. Opt.* **50** 886–90

[29] Bai H, Shan M, Zhong Z, Guo L and Zhang Y 2015 Common path interferometer based on the modified Michelson configuration using a reflective grating *Opt. Laser Eng.* **75** 1–4

[30] Anand A, Chhaniwal V, Mahajan S, Trivedi V, Singh A, Leitgeb R and Javidi B 2014 Self-referencing digital holographic microscope for dynamic imaging of living cells *Proc. SPIE* **9117** 91170X

[31] Singh S G, Anand A, Leitgeb R A and Javidi B 2012 Lateral shearing digital holographic imaging of small biological specimens *Opt. Express* **20** 23617–22

[32] Singh V, Joshi R, Tayal S and Mehta D S 2019 Speckle-free common-path quantitative phase imaging with high temporal phase stability using a partially spatially coherent multi-spectral light source *Laser Phys. Lett.* **16** 025601

[33] Singh V, Tayal S and Mehta D S 2018 Highly stable wide-field common path digital holographic microscope based on a Fresnel biprism interferometer *OSA Contin.* **1** 48–55

[34] Singh V, Tayal S and Mehta D S 2019 Single shot fringe projection profilometry using tunable frequency Fresnel biprism interferometer for large range of measurement' *Opt. Commun.* **451** 104–10

[35] Edwards C, Bhaduri B, Nguyen T, Griffin B G, Pham H, Kim T, Popescu G and Goddard L L 2014 Effects of spatial coherence in diffraction phase microscopy *Opt. Express* **22** 5133–46

[36] Soto J M, Rodrigo J A and Alieva T 2018 Optical diffraction tomography with fully and partially coherent illumination in high numerical aperture label-free microscopy *Appl. Opt.* **57** A205

[37] Rodrigo J A and Alieva T 2014 Rapid quantitative phase imaging for partially coherent light microscopy *Opt. Express* **22** 13472–83

[38] Rodrigo J A and Alieva T 2014 Illumination coherence engineering and quantitative phase imaging *Opt. Lett.* **39** 5634–7

[39] Nguyen T H, Edwards C, Goddard L L and Popescu G 2014 Quantitative phase imaging with partially coherent illumination *Opt. Lett.* **39** 5511–4

[40] Bianco V, Memmolo P, Leo M, Montresor S, Distante C, Paturzo M, Picart P, Javidi B and Ferraro P 2018 Strategies for reducing speckle noise in digital holography *Light: Sci. Appl.* **7** 48

Chapter 4

Structured illumination quantitative phase microscopy

This chapter describes a super-resolution technique in which the specimen is illuminated with a structured interference pattern in order to increase the resolution of the system beyond diffraction limit. The chapter discusses various illumination strategies for generating the structured light pattern. It details the off-axis as well as common-path optical configurations to achieve structured light phase microscopy. Phase reconstruction algorithms shows the increase in resolution with a gain factor of 2. Structured illumination microscopy (SIM) finds its future in biomedical applications where the information about the samples is in nanometric scale.

4.1 Diffraction-limited resolution of phase microscopy

Optical microscopes are used to visualize micron size objects but they have diffraction-limited resolution due to the wave nature of light. The diffraction limit of resolution was first introduced by Ernst Abbe in 1873, which is termed as $\frac{\lambda}{2NA}$, i.e., it is proportional to the wavelength of light, and inversely proportional to the numerical aperture (NA) of the objective lens [1]. The maximum value of NA for a dry objective lens is 1.0 (for an oil immersed objective lens the value could be 1.44), thus the smallest object size that could be resolved is $\geqslant \frac{\lambda}{2}$. Further, the two-point objects cannot be differentiated if the lateral distance between them is less than the above diffraction limit of the system, as shown in figure 4.1. In this figure the optical system could be optical microscopes (bright-field, dark-field, epi-fluorescence, confocal, phase contrast, differential interference contrast, and quantitative phase), spectrometers, spectroscopes, telescopes etc. All these instruments have diffraction-limited resolution.

There are various subcellular structures that are smaller than the diffraction-limited resolution of optical microscopes. To image these subcellular structures, in the 1990s, super-resolution microscopes, also known as nano-scopes, were developed, which can resolve beyond the diffraction limit of the system [2, 3]. Stimulated

doi:10.1088/978-0-7503-3987-2ch4

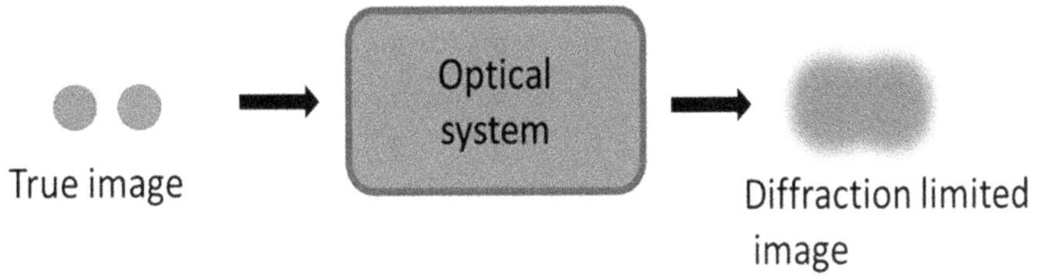

Figure 4.1. Diffraction-limited image of the two-point objects.

emission depletion (STED) microscopy [4, 5], photo-activated localization microscopy (PALM) and stochastic optical reconstruction microscopy (STORM) [6, 7] and direct stochastic optical reconstruction microscopy (dSTORM) are super-resolution techniques that are implemented mainly in fluorescence imaging, providing resolution much beyond the diffraction limit of the system, i.e., up to single molecule level ~20–50 nm. But most of these systems are bulky, costly and computationally heavy. SIM [8–13] is another super-resolution technique which illuminates the sample with a structured interference pattern to extend the resolution of the system beyond the diffraction limit. SIM is much simpler and less cumbersome. All the aforementioned super-resolution techniques are fluorescence based techniques and require exogenous contrast agents for labelling cells and tissues. But for many biological applications, label-free super-resolution imaging with quantitative information is required. Recently, a few techniques have been developed for super-resolution quantitative phase imaging using structured illumination quantitative phase microscope (SIQPM) [14–16], and Fourier ptychography [17, 18]. By means of using these techniques one can obtain both super-resolution as well as quantitative information along with wide field of view. In this chapter we describe various schemes which have been implemented in SIQPM.

4.2 Structured illumination schemes for phase microscopy

Figures 4.2(a) and (b) show a schematic diagram of off-axis QPM systems with conventional illumination and structured illumination. In figure 4.2(b) the sample is illuminated using two beams which form an interference pattern on the sample. The structured illumination is based on the phenomenon of the moiré beating [8–13]. When two high frequency grating patterns are superimposed at different angles, as a result, a low fringe density pattern generates, which is termed as a moiré pattern, as shown in figure 4.2. To obtain the high spatial frequency components of the object, the sample is illuminated with the structured pattern to overcome the limitation present due to optical diffraction. In the diffraction-limited system, the optical transfer function (OTF) pass band can gather only diffraction-limited information, also known as cut-off frequency of the system, which is shown in figure 4.3(a). The high frequency information beyond the cut-off frequency is lost. In SIM, the object is illuminated with the structured pattern, which collects the high frequency components of the sample by forming a moiré fringe pattern. Illumination of the

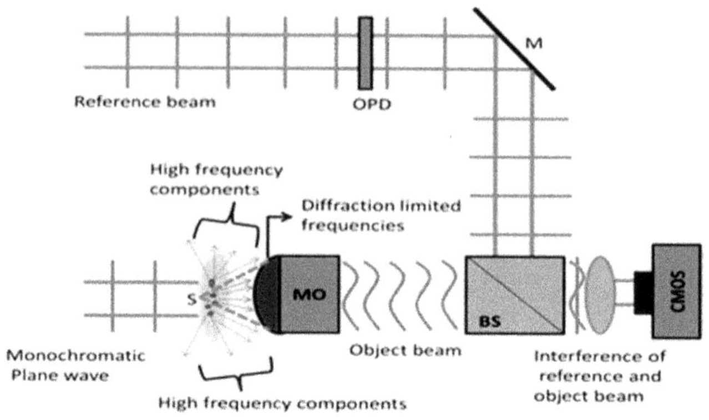

(a)

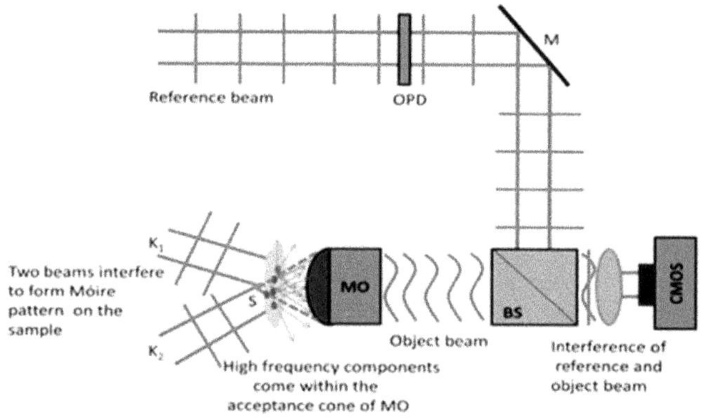

(b)

Figure 4.2. (a) Conventional configuration of QPM, (b) illustrating the generation of low frequency moiré fringes by means of superposition of two grating patterns.

object with the changing moiré pattern leads to the collection of higher spatial frequency components from the object and hence the resolution is significantly improved. In this process the maximum amount of higher spatial frequency components are collected and hence the bandwidth of the microscope is increased.

SIM can work both in reflection as well as transmission mode. The frequency domain for SIM is shown in figure 4.3(b), which represents the extended frequency domain, thereby resulting in a highly resolved image. In SIM, the object is illuminated with the structured pattern. The image of the sample can be mathematically represented as

$$g(x, y) = p_c(x, y) \otimes \left[t(x, y) * (p_c(x, y) \otimes s(x, y)) \right] \tag{4.1}$$

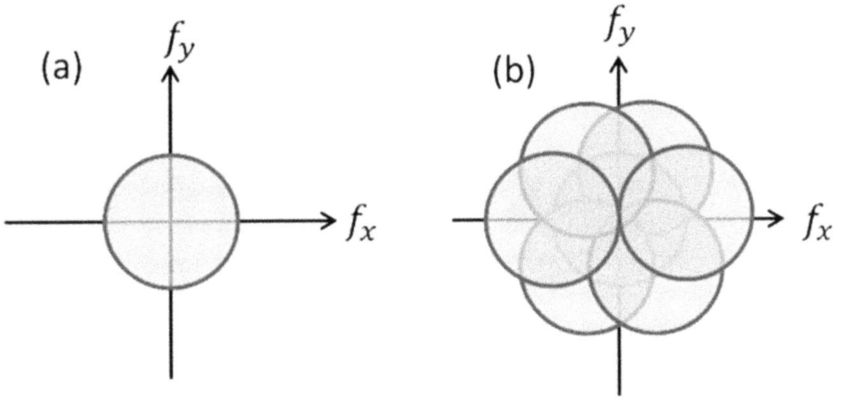

Figure 4.3. (a) Frequency domain of the diffraction-limited system. (b) Frequency domain of the structured illuminated system.

where $p_c(x, y)$ is the point spread function, $t(x, y)$ is the transmittance function of the object and $s(x,y)$ is the illumination pattern which is mathematically represented as $s(x, y) = \cos(2\pi f_o x + \varphi_n)$.

In the frequency domain, the spatial frequency components of light scattered from the sample are the convolution of the object and the structured illumination frequency spectrum, low pass filtered with the system's transfer function [8–13]. The system's transfer function is the OTF or coherent transfer function (CTF) in fluorescence or coherent imaging system, respectively [19]. In fluorescence imaging, the sample distribution, illumination pattern and the OTF are in the intensity domain, whereas in the coherent imaging system, all the terms including object distribution, illumination pattern distribution as well as the CTF of the system are in the imaginary domain, thus it contains the imaginary components, thus they are mathematically complex. There are various schemes for the structured illumination phase microscope. The lateral resolution of SIM is improved due to increase in the $NA_{\text{illumination}}$, and can be given by the formula $\Delta x = \frac{1.22\lambda_{\text{ex}} n}{(NA_{\text{det}} + NA_{\text{illumination}})}$. Since NA_{det} is fixed, the lateral resolution can therefore be improved by means increasing the $NA_{\text{illumination}}$, which can be increased by collecting the maximum number of spatial frequency components, i.e., $2NA_{\text{illumination}}/\lambda_{\text{ex}}$. This is possible with the help of structured illumination.

4.2.1 Off-axis geometry for SIQPM

Off-axis SIQPM based on a Mach–Zehnder interferometer system in transmission mode is shown in figure 4.4. In this set-up light from the laser is spatially filtered using MO and pinhole, and lens L collimates the beam which is further split into the reference and the object beam using the beam splitter. In the object arm, a diffraction grating is used which splits the beam into multiple diffraction orders and those pass via a 4f system (L_1 to OBJ). In this path the DC term is blocked so that only ±1 diffraction orders are interfered to illuminate the sample with the

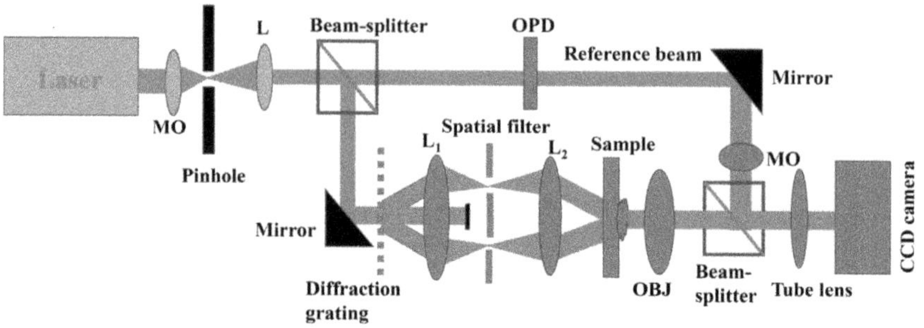

Figure 4.4. Optical configuration for the off-axis Mach–Zehnder based structured illumination phase microscope (L, L1, L2 and L3—lenses, BS—beam splitter, OPD—optical path delay element, M—mirror, DG—diffraction grating, OBJ—microscopic objective lens).

structured pattern. Good contrast of interference pattern is observed because laser light is spatially and temporally highly coherent and the magnitude of the interfering beams ±1 diffraction orders are equal. The interference pattern illuminating the sample can be phase shifted (changing pattern) by means of translating the grating linearly using a motorized stage. A minimum of four phase-shifted interferograms with phase shift of $\pi/2$ each can be used for illuminating the sample. For each illuminating interferogram the objective lens collects the high spatial frequency components of the object beam. The second 4f system OBJ to L_2 is used to coherently image the transmitted diffraction pattern from the object.

Finally, this coherent light image is interfered with an off-axis reference wave to obtain amplitude and phase of the object distribution. To compensate the optical path in the reference arm, the optical path length delay element is added in the reference arm. The results of isolated mesenchymal stem cells reported in [14] using the off-axis geometry Mach–Zehnder type SIQPM are found to have improved resolution. The samples were prepared using the cell cultured technique mentioned in [14]. Experimental results of diffraction-limited and high-resolution phase images are compared, which represent the heterogeneous distribution of the condensed nucleus material. The high contrast and intricate nuclear condensate with fine features are visible in a high resolution image recorded using SIQPM [14]. All three zoom regions (A, B and C) also indicate the improvement in the image quality with SIQPM compared to wide field (WF).

In general, SIM is implemented by means of illuminating the sample with phase-shifting interference patterns. This requires a minimum of four phase shifted interferograms to be projected onto the sample, therefore, the experimental set-up becomes a bit costly. Recently, phase-shifting-free structured illumination digital holography microscopy (DHM) has been implemented for the enhancement of resolution [15]. In addition, angular multiplexing along with SI was also implemented to further improve the resolution in DHM, in which two reference beams with different carrier frequencies and orthogonal polarization states are used. Figure 4.5 shows a schematic diagram of SI-DHM with angular multiplexing which is single-shot in nature. More details can be found in reference [15]. In this scheme a diffraction grating, collimating lens and two polarizing filters were used to generate two beams with orthogonal polarization, as shown in figure 4.5 [15].

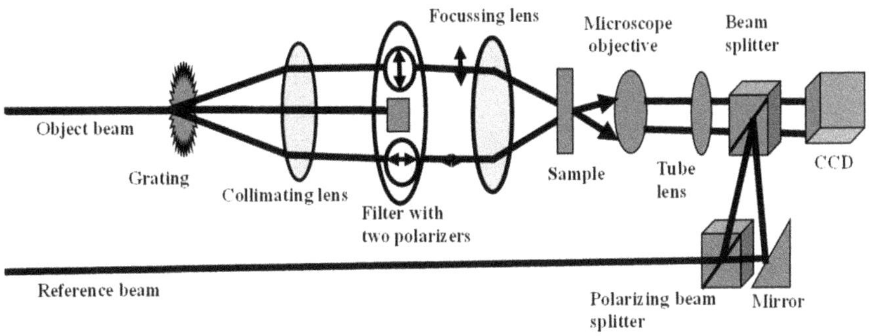

Figure 4.5. Illustrating the single-shot SI-DHM with angular multiplexing.

The two beams are focussed onto the sample and a combination of objective lens and tube lenses were used to form the microscopic images onto the CCD camera. The reference beam passes through the polarizing beam splitter to separate two orthogonal polarization components and finally both the beams are made incident onto the beam splitter to interfere with the objects beams. With this configuration multi-plex interferograms will be generated corresponding to each polarization component. Figure 4.6 shows the comparison of results with and without SI [15]. Significant improvement in resolution was obtained.

4.3 Common-path geometries for SIQPM

4.3.1 Structured illumination lateral shearing interferometry based SIQPM

A schematic diagram of common-path lateral shearing interferometer based SIQPM is shown in figure 4.7. In [16], the authors used a He–Ne laser light source to illuminate a grating. The diffracted orders of light from the grating are used to create the illumination pattern onto the sample using the 4f system L1 to L2. The MO is used to collect the sample amplitude distribution and imaged onto the sensor plane. S_1 and S_2 are two virtual point sources to generate interference between the two wavefronts. The authors in [16], had recorded two images to separate the low and high spatial frequency components mixed in SI and recover the high-resolution image without the need of multiple phase shifted images. To accomplish this job, they have introduced two apertures in the system.

They first recorded the low frequency image by partially closing aperture 1 which blocks the ± 1 orders and opening aperture 2 to allow the DC components to pass through it. The second image is the high frequency image which was recorded by blocking aperture 2 and opening aperture 1, which allows the ± 1 orders and results in a high frequency image. Further, lenses L1 and L2 were placed in such a way that the object plane and the grating are conjugate planes.

They placed L1 lens at 34.50 mm from the grating, and the distance between L1 and L2 lenses is 31.00 mm. The L1 and L2 lenses have focal length of 50.20 mm and 62.90 mm, respectively. The common-path geometry benefits by increasing the temporal phase stability of the system. The results on Selenastrum cells, which are sickle-shaped green algae, imaged by O'Connor *et al* [16] using the type common-path sheared based

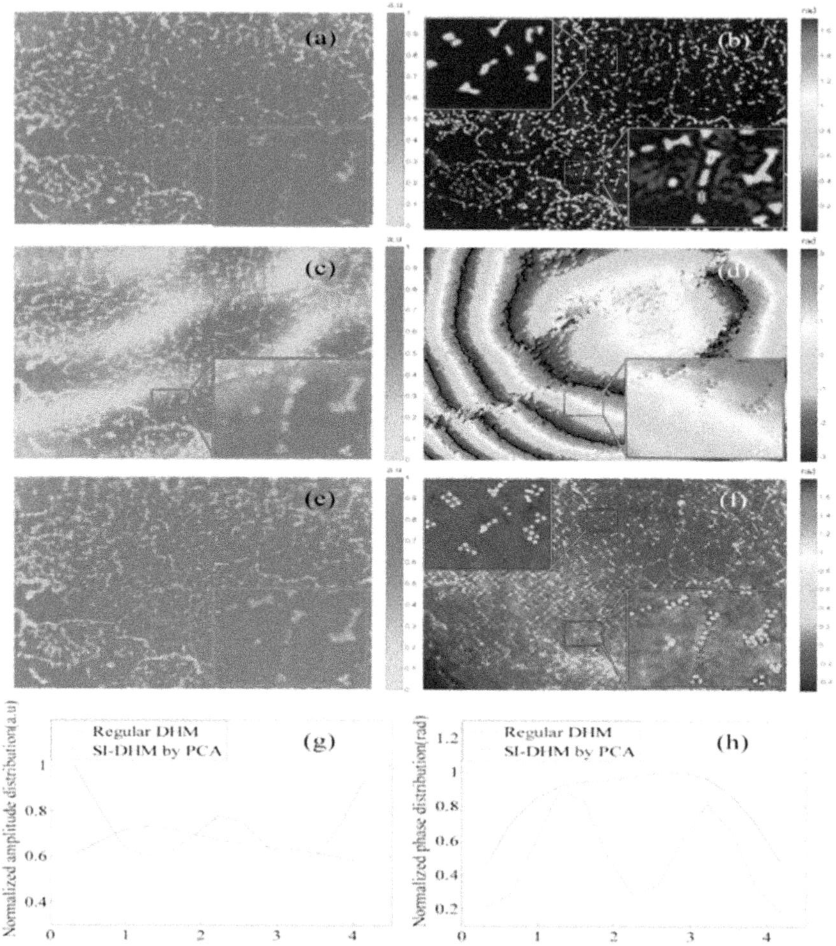

Figure 4.6. (a) Amplitude images polystyrene spheres using DHM and (c) SI-DHM. (b) Phase image of specimen using DHM, and (d) phase image of specimen using SI-DHM; (e) and (f) are the resolution enhanced images using SI-DHM, and (g) and (h) are the comparison resolution improvement with and without SI. Reproduced with permission from reference [15], copyright 2018 Optica Publishing Group. All Rights Reserved.

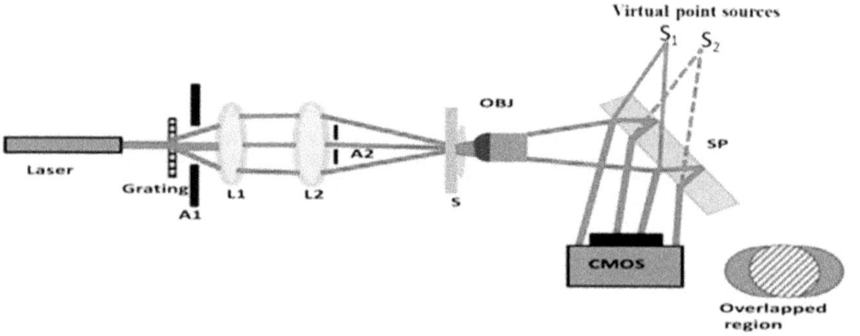

Figure 4.7. Optical configuration for a common-path sheared based SIPM. (A1, A2—aperture, L1, L2—lenses, S—sample, OBJ—microscope objective lens, SP—sheared plate). Replotted from reference [16].

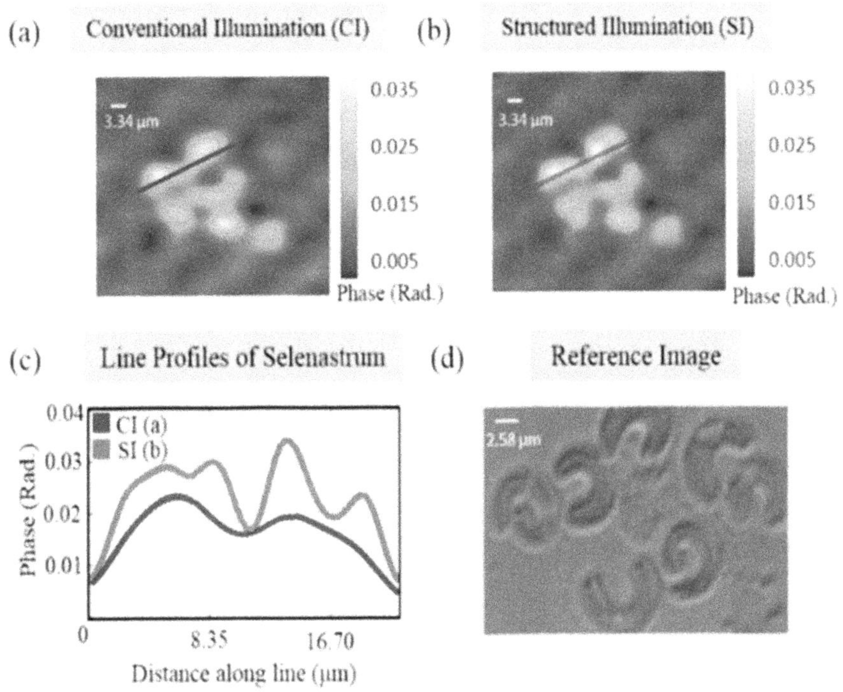

Figure 4.8. 2D pseudo-color phase reconstruction for (a) conventional illumination (CI), and (b) structured illumination (SI) of Selenastrum. (c) Line profiles for both cases of Selenastrum; CI shown in blue and SI in red. (d) Light microscope image of Selenastrum (60 × /0.85 NA). Reproduced with permission from reference [16], copyright 2019 Optica Publishing Group. All Rights Reserved.

SIPM are shown in figure 4.8. The conventional and structured illumination phase images are shown in figures 4.8(a) and (b). The line profile shown in figure 4.8(c) corresponding to blue and red line represents that there is a clear improvement in the resolution when the object is illuminated with the structured pattern. Figure 4.8(d) represents the reference image of the same cell imaged by light microscope.

4.3.2 Structured illumination using Fresnel bi-prism interferometer

The SI patterns are also generated using a Fresnel-bi-prism interferometer system [17]. By means of placing a bi-prism in one of the arms of the DHM [17] the two interfering beams are generated and focused onto the sample. These two beams mutually interfere on the sample plane. Further, the beams are of equal intensity generated by the prism and hence are efficient. In addition, by means of displacing or rotating the bi-prism one can generate phase-shifting and rotating interference patterns required for SI. About a two-fold improvement in spatial resolution was achieved using this method. Figure 4.9 shows SI-DHM using a Fresnel bi-prism for generating interference patterns for illuminating the sample. The two beams that are of equal intensity interfere onto the sample and the scattered light collected by the MO lens is further interfered with the reference beam, thus generating the SI-DHM. Figure 4.10 shows the experimental results with normal DHM and with SI-DHM.

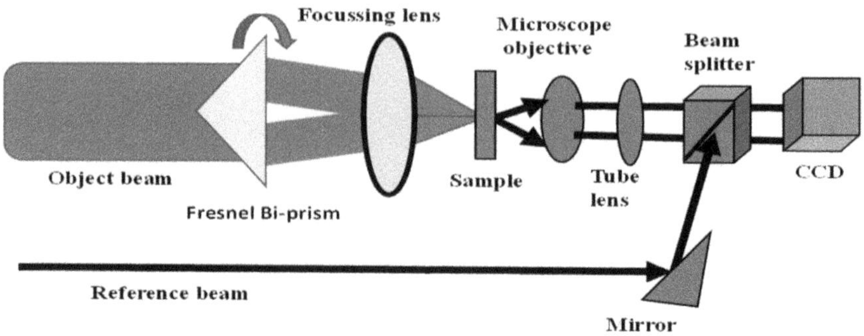

Figure 4.9. Illustrating SI-DHM using a bi-prism interferometer.

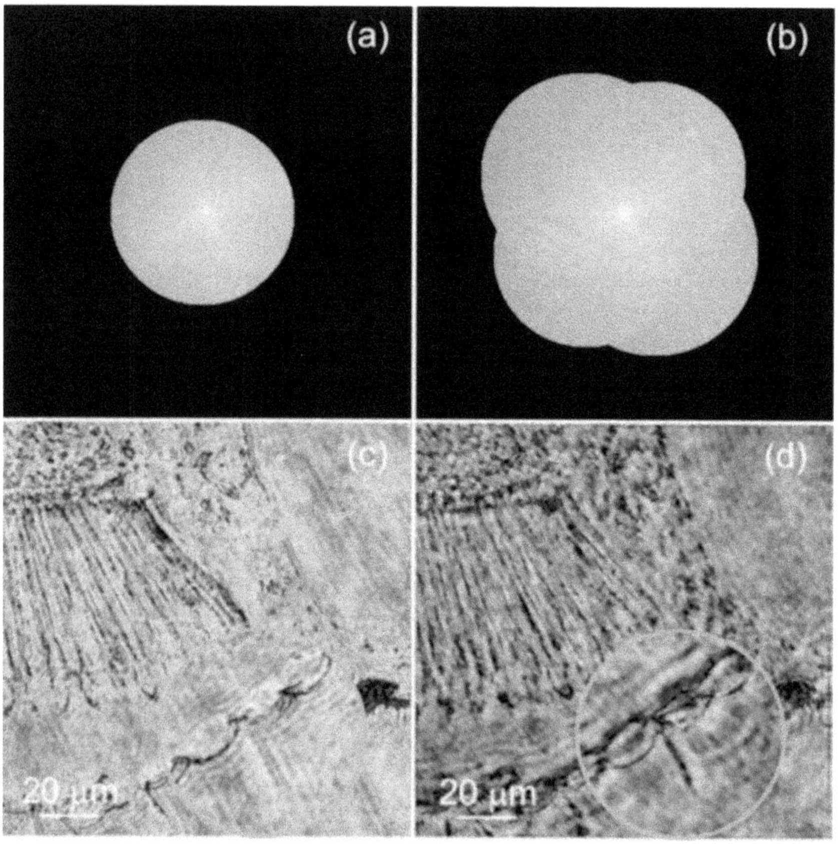

Figure 4.10. (a) and (b) Coherence transfer functions of normal DHM and SI-DHM, respectively, and (c) and (d) are the amplitude contrast images section of the head of a drosophila melanogaster fly with normal DHM and SI-DHM, respectively. Reproduced with permission from reference [17], copyright 2014 Optica Publishing Group. All Rights Reserved.

4.3.3 Structured illumination diffraction phase microscopy

In [18], the authors designed a common-path SI phase microscope using a diffraction grating, as shown in figure 4.11, and named it as structured illumination diffraction

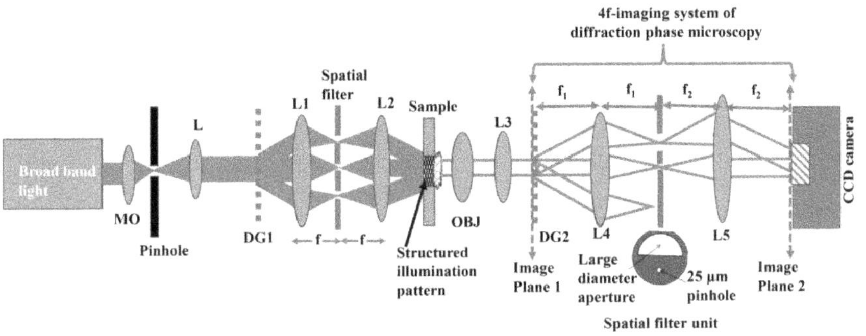

Figure 4.11. Optical configuration for the common-path diffraction grating based SIPM (L, L1, L2, L3, L4 and L5-lenses, OBJ—microscope objective lens, DG1, DG2—diffraction gratings, SMF—single-mode fibre). Replotted from reference [18].

phase microscope (SI-DPM). A broad-band light using the super-continuum light source was used for illumination to improve the spatial phase sensitivity of the system. Light from the source first spatially filtered using microscope objective (MO) and a pinhole and is further collimated using lens (L). The collimated beam passes through a diffraction grating (DG1). The ±1 and 0 orders from DG1 (50 lpmm) were used for the SI onto the sample via 4f lens system L1 to OBJ. The sample illuminated with the pattern is then imaged via a second 4f lens system OBJ to L2 onto a second diffraction grating DG2 (100 lpmm). Then the third 4f lens system L3 to L4 was designed where a pinhole will spatially filter the 0th order, allow +1 order to pass which carries the sample information and block all other orders. Here, the common-path geometry allows the high temporal phase stability and also broad-band light source reduces the speckle noise and allows the high spatial phase sensitivity and is advantageous for allowing the broad-band illumination phase reconstruction over the large field of view.

The results of a single mesenchymal stem cell under broad band illumination are reported in [18] using the type common-path diffraction based SIPM. The diffraction-limited images are compared to sub-diffraction resolution images and improvement in resolution. The improvement in the resolution of phase images can be clearly observed in the images reported using SI [18]. This system is better as compared to the off-axis type interferometer as it increases the temporal phase stability and it also improves the phase measurement accuracy as it utilizes a broad- band light source which is speckle-free and does not generate spurious fringes which are responsible for the degradation of the results. Thus, the common-path and speckle-free light source along with SI could be used for the precise and highly resolved phase measurement of the specimen.

4.4 Phase reconstruction algorithms for SIPM

In conventional SIM, fluorescence imaging is performed which is incoherent in nature. In incoherent imaging, there is a linear relationship between the sample distribution and the intensity at the image plane. Thus, the intensity at the image plane is the linear transform of the object distribution. In contrast to fluorescence SIM, i.e., in the case of SIPM, the coherent imaging is performed which has a linear relationship between the object transmittance function and the image field. Thus, the image field is a linear transform of the object transmittance function. Let us consider the object complex transmittance function to be $t(x, y)$, the structured illumination field set by two interfering beams to be $s(x, y) = \cos(2\pi f_o x + \varphi_n)$, where $|f_o| \leqslant f_c$, f_o is the illumination's frequency vector and f_c is the cut- off frequency, k is the direction vector. The coherent image formed is $g(x, y) = p_c(x, y) \otimes \left[t(x, y)^*(p_c(x, y) \otimes s(x, y))\right]$ which is previously represented by equation (4.1).

This image is then interfered by an off-axis reference wave $(x, y)= e^{-j(k_x x+k_y y)}$, considered without proportionality constant for simplicity in [14, 19–21]. The mathematical presentation of the interference pattern on the camera is

$$I(x, y) = |g(x, y) + R(x, y)|^2$$

$$=1 + |g(x, y)|^2 + g(x, y)e^{-j(k_x x+k_y y)} + g^*(x, y)e^{j(k_x x+k_y y)} \tag{4.2}$$

From the Fourier method, the complex carrier frequency information can be completely reconstructed and the phase and amplitude information can be obtained. By taking Fourier transform of equation (4.1), the equation can be written as [22]

$$G(f_x, f_y) = P_c(f_x, f_y) * \left[T(f_x, f_y) \otimes \left(P_c(f_x, f_y) * S(f_x, f_y)\right)\right] \tag{4.3}$$

$G(f_x, f_y)$, $P_c(f_x, f_y)$, $T(f_x, f_y)$, $S(f_x, f_y)$ are the Fourier transform of $g(x, y)$, $p_c(x, y)$, $t(x, y)$ and $s(x, y)$, respectively. The frequency response of the diffraction-limited coherence system can be defined using the amplitude transfer function [20]. From the coherent theory, $P_c(f_x, f_y)$, acts as the tophat filter which allows all the frequencies within the cut-off of the diffraction band and sharply rejects all the frequencies beyond the diffraction limit of the system [19–21]. By taking the Fourier transform of equation (4.3), the following equation can be obtained [14, 19–21]

$$G_n(f_x, f_y) = P_c(f_x, f_y) * \left[T(f_x - f_0, f_y)e^{-j\varnothing_k} + T(f_x + f_0, f_y)e^{j\varnothing_k}\right] \tag{4.4}$$

In the case of conventional SIM, by phase shifting $s(x, y)$, one can linearly solve for $T(f_x - f_0, f_y)$ and $T(f_x + f_0, f_y)$ and place them back to their correct position in Fourier space to retrieve the image, with effective system's transfer function $P_{\text{eff}}(f_x, f_y) = P_c(f_x - f_0, f_y) + P_c(f_x + f_0, f_y)$. By rotating the illumination pattern, one can fill the gap in Fourier space, and maximum resolution achieved is when $|f_o| = f_c$ and corresponds to a gain factor of 2. From equation (4.4), the increment in the resolution can be theoretically achieved by taking only two rotations. However, in [14], the authors had taken six raw acquisitions for better signal-to-noise ratio and fidelity of reconstruction. In summary SIM phase microscopy has a great future for biomedical applications. If SI-QPM is implemented in common-path configurations along with spatially partially coherent light then a quantitative phase nanoscopy can be realized, leading to information about samples in nanometric scale.

References

[1] Saleh B E and Teich M C 1991 *Fundamentals of Photonics* (New York: Wiley)

[2] Schermelleh L, Heintzmann R and Leonhardt H 2010 A guide to super-resolution fluorescence microscopy *J. Cell Biol.* **190** 165–75

[3] Huang B, Babcock H and Zhuang X 2010 Breaking the diffraction barrier: super-resolution imaging of cells *Cell* **143** 1047–58

[4] Hein B, Willig K I and Hell S W 2008 Stimulated emission depletion (STED) nanoscopy of a fluorescent protein-labeled organelle inside a living cell *Proc. Natl Acad. Sci.* **105** 14271–6

[5] Hell S W and Wichmann J 1994 Breaking the diffraction resolution limit by stimulated emission: stimulated-emission-depletion fluorescence microscopy *Opt. Lett.* **19** 780–2

[6] Betzig E *et al* 2006 Imaging intracellular fluorescent proteins at nanometer resolution *Science* **313** 1642–5

[7] Bates M, Huang B and Zhuang X 2008 Super-resolution microscopy by nanoscale localization of photo-switchable fluorescent probes *Curr. Opin. Chem. Biol.* **12** 505–14

[8] Gustafsson M G 2000 Surpassing the lateral resolution limit by a factor of two using structured illumination microscopy *J. Microsc.* **198** 82–7

[9] Gustafsson M G 2005 Nonlinear structured-illumination microscopy: wide-field fluorescence imaging with theoretically unlimited resolution *PNAS* **102** 13081–6

[10] Schermelleh L *et al* 2008 Sub-diffraction multi-color imaging of the nuclear periphery with 3D structured illumination microscopy *Science* **320** 1332–6

[11] Kner P, Chhun B B, Griffis E R, Winoto L and Gustafsson M G 2009 Super-resolution video microscopy of live cells by structured illumination *Nat. Methods* **6** 339–42

[12] Shao L, Kner P, Rego E H and Gustafsson M G 2011 Super-resolution 3D microscopy of live whole cells using structured illumination *Nat. Methods* **8** 1044–6

[13] Fiolka R, Shao L, Rego E H, Davidson M W and Gustafsson M G 2012 Time-lapse two-color 3D imaging of live cells with doubled resolution using structured illumination *Proc. Natl Acad. Sci.* **109** 5311–5

[14] Chowdhury S and Izatt J 2013 Structured illumination quantitative phase microscopy for enhanced resolution amplitude and phase imaging *Biomed. Opt. Express* **4** 1795–805

[15] Li S, Ma J, Chang C, Nie S, Feng S and Yuan Caojin 2018 Phase-shifting-free resolution enhancement in digital holographic microscopy under structured illumination *Opt. Express* **26** 23572

[16] O'Connor T, Doblas A and Javidi B 2019 Structured illumination in compact and field-portable 3D-printed shearing digital holographic microscopy for resolution enhancement *Opt. Lett.* **44** 2326–29

[17] S.-Ortiga E, Martínez-Corral M, Saavedra G and Garcia-Sucerquia J 2014 Enhancing spatial resolution in digital holographic microscopy by biprism structured illumination *Opt. Lett.* **39** 2086

[18] Chowdhury S and Izatt J 2014 Structured illumination diffraction phase microscopy for broadband, sub-diffraction resolution, quantitative phase imaging *Opt. Lett.* **39** 1015–8

[19] Ma Y, Wen K, Liu M, Zheng J, Chu K, Smith Z J, Liu L and Gao P 2021 Recent advances in structured illumination microscopy *J. Phys.: Photon.* **3** 024009

[20] Tian L, Li X, Ramchandran K and Waller L 2014 Multiplexed coded illumination for Fourier Ptychography with an LED array microscope *Biomed. Opt. Express* **5** 2376–89

[21] Zheng G, Horstmeyer R and Yang C 2013 Wide-field, high-resolution Fourier ptychographic microscopy *Nat. Photonics* **7** 739–45

[22] Goodman J W 2005 *Introduction to Fourier Optics* (Englewood, CO: Roberts & Company)

Chapter 5

Multimodal on-chip nanoscopy and quantitative phase microscopy

In this chapter, we present a multimodal fluorescence imaging and quantitative phase microscopy platform to obtain comprehensive information of the object. In fluorescence imaging, we review fundamental principle of fluorescence imaging, total internal reflection fluorescence imaging and single-molecule localization microscopy. A unique approach, i.e., on-chip nanoscopy and its integration with partially spatially coherent quantitative phase microscopy is presented in this chapter. On-chip nanoscopy is useful to image features of size 50 nm in lateral direction while phase microscopy can be utilized for highly sensitive axial measurements and, therefore, a multimodal platform will be useful in biological applications.

5.1 Fluorescence imaging

Although, quantitative phase microscopy (QPM) techniques are capable of quantifying several biophysical parameters, and can provide accumulative information of all the individual components such as cell nucleus, membrane and mitochondria etc, multimodality has the advantage of obtaining comprehensive and complementary information about the specimen by combining different optical imaging modalities. To image different functions of the cells with enhanced contrast, fluorescence imaging is widely adopted by the scientific community [1]. In fluorescence imaging, a labelling agent (fluorophore) is used to tag with the specimen in order to generate high contrast images. Because of its intrinsic selectivity, fluorescence imaging has become the mainstay of microscopy in biomedical imaging applications [1, 2]. Alternatively, intrinsic fluorescence from biological samples known as auto-fluorescence also can be used in some cases.

The availability of a large number of fluorophores covers large spectral range which allows simultaneous imaging of different cellular, sub-cellular and intracellular components [2, 3]. Fluorescence is a well-known phenomenon which involves absorption and emission of a small range of light wavelengths by a

doi:10.1088/978-0-7503-3987-2ch5

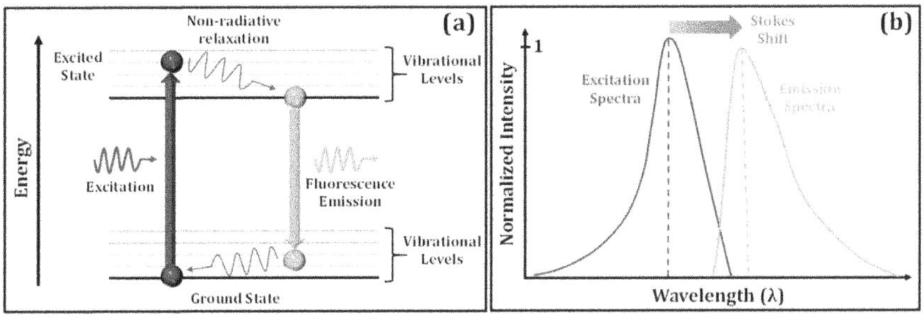

Figure 5.1. Fluorescence phenomenon and Stokes shift are shown here. (a) Energy diagram of a fluorophore excitation and emission. (b) Stokes shift in fluorophore excitation and emission spectra [4].

fluorophore. These stained specimens are illuminated with a light of specific wavelength which is absorbed by the fluorophores and the emission is observed typically at larger wavelengths, as shown in figure 5.1. This emission occurs within nanoseconds after the absorption and the difference between absorption and emission wavelengths is known as Stokes shift. This shift is the most crucial property which makes fluorescence microscopy very effective to enhance the contrast by completely removing the illumination background.

The energy level diagram of a fluorescence material in figure 5.1(a) shows the excitation of electrons by shining with photons having high energy than the band gap of the molecules. The decay of electrons back to the ground level emits the fluorescent signal which produces a high contrast as compared to the background. Depending on the type of applications and resolution, several fluorescence microscopic techniques were proposed such as (i) epi-fluorescence [5], (ii) total internal reflection fluorescence [6] (TIRF) and (iii) super-resolutions techniques [7] etc.

5.1.1 Epi-fluorescence imaging

Epi-fluorescence is the most common technique to enhance the contrast in an imaging system. In an epi-fluorescence system, a high power light source shines through an excitation filter having a very narrow region of wavelength. The excitation filter selects only a specific wavelength in order to excite the specific fluorophore of interest. The object is illuminated by the incident light through a microscope objective (MO) and the fluorescent signal is collected by the MO, as shown in figure 5.2. The incident light is reflected off by using the emission filter so that only the fluorescent signal reaches the camera plane. In this scheme the signal-to-noise ratio (SNR) is poor as the illumination extends to the entire specimen thus generating fluorescence signal from the entire volume of the specimen [8]. Excitation and emission photons beyond the depth of focus of the objective lens contribute the background noise and therefore degrade the quality of the final image. The direct illumination may also lead to the photo-bleach and light stress on living cells due to high intensity of incident light.

5.2 Total internal reflections fluorescence microscopy

There are various schemes which were proposed to reject out-of-focus light and improve the SNR of fluorescent images. Total internal reflections fluorescence (TIRF)

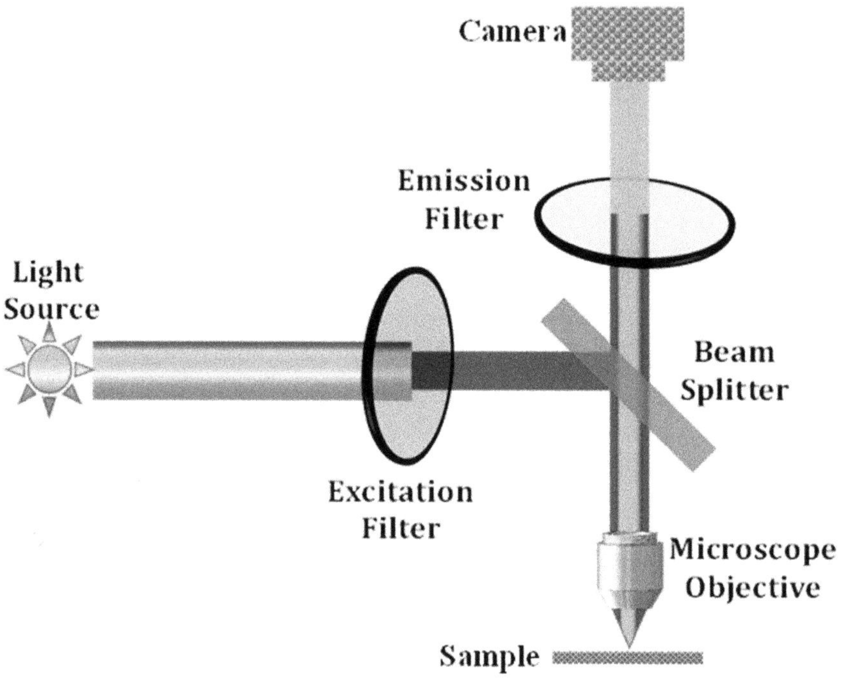

Figure 5.2. Schematic diagram of epi-fluorescence microscopy set-up.

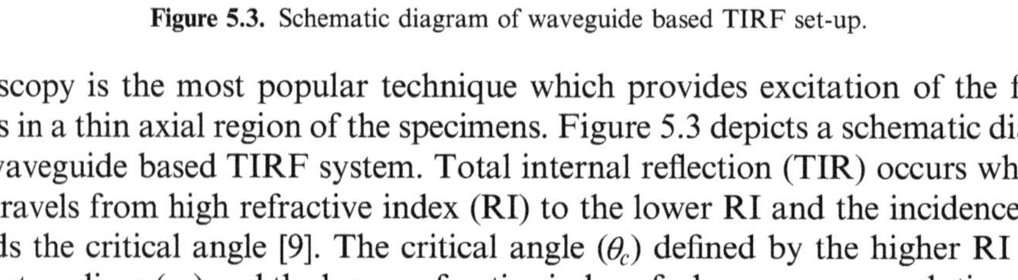

Figure 5.3. Schematic diagram of waveguide based TIRF set-up.

microscopy is the most popular technique which provides excitation of the fluorophores in a thin axial region of the specimens. Figure 5.3 depicts a schematic diagram of a waveguide based TIRF system. Total internal reflection (TIR) occurs when the light travels from high refractive index (RI) to the lower RI and the incidence angle exceeds the critical angle [9]. The critical angle (θ_c) defined by the higher RI of the incident medium (n_1) and the lower refractive index of a homogeneous solution (n_2) is:

$$\theta_c = \sin^{-1}(n_2/n_1) \tag{5.1}$$

In the case of a waveguide, an evanescent field is generated by total internal reflection of the light at the interface. The evanescent field intensity decays exponentially in the z direction and can be expressed as [10]:

$$I_e(z) = I_0 \exp(-z/d) \tag{5.2}$$

where d is the characteristic depth of penetration and I_0 is the incident intensity at the interface. If λ is the wavelength of incident light, depth of penetration can be defined as:

$$d = \frac{\lambda}{4\pi\sqrt{n_1^2 \sin^2\theta - n_2^2}} \tag{5.3}$$

The sample of interest is kept on the top of the surface where evanescent field is generated, and the specimen is illuminated by this, as shown in figure 5.3. This scheme provides a high-resolution image with an excellent SNR, less light bleaching and less light stress.

5.3 Super-resolution fluorescence imaging

The above-mentioned fluorescence techniques suffer from diffraction limited resolution of the microscopic system. Far field optical nanoscopy is a prominent tool to resolve sub-cellular structures in a biological specimen by means of breaking the optical diffraction limit. Optical nanoscopy includes a variety of innovative techniques, such as, stimulated emission depletion (STED) microscopy [11], structured illumination microscopy (SIM) [12], photo-activated localization microscopy (PALM) [13], super-resolution optical fluctuation imaging (SOFI) [14] and direct stochastic optical reconstruction microscopy (dSTORM) [15]. These techniques offer super-resolution (<200 nm) imaging of sub-cellular structures, of biological specimens such as cell membrane, fine boundaries, proteins, lipids, and nucleic acids by specific fluorescence tagging [16, 17]. Each technique has advantages and disadvantages compared to the others.

SIM has recently gained popularity in various applications such as live-cell imaging due to fast image acquisition time [18]. On the other hand, dSTORM offers extremely high resolution (below 50 nm) imaging in living cells [19], characterization of proteins [20] and imaging actin filaments [21]. dSTORM is based on the principle of single-molecule localization microscopy (SMLM). Despite very high resolution, complexity and affordability of the conventional dSTORM techniques makes it impractical for a wide range of biological applications.

5.3.1 On-chip nanoscopy (single-molecule localization microscopy technique)

A chip based nanoscopy system was recently proposed to reduce the complexity of conventional dSTORM [22]. In chip based nanoscopy, a waveguide chip is used for the illumination rather than through the imaging lens. An evanescent field is used to generate the excitation of single molecules on top of waveguides, which makes it possible to image arbitrarily large areas compared to conventional dSTORM [15]. Secondly, decoupling excitation and collection path enables miniaturization of the overall system. Chip based nanoscopy recently showed its capability of imaging a large field of view, i.e., 0.5 × 0.5 mm with 70 nm spatial resolution [23].

5.3.2 Working principle of on-chip nanoscopy

In chip based nanoscopy, the sample is prepared on top of a photonic integrated circuit (PIC), as shown in figure 5.4. The PIC structure was fabricated onto a

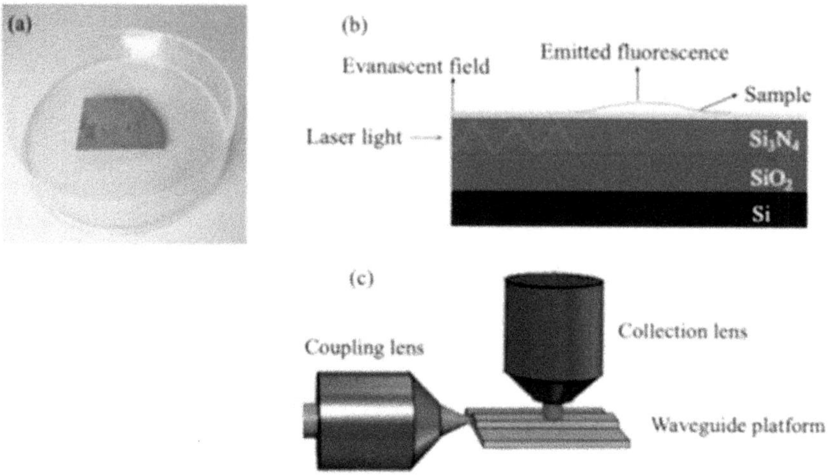

Figure 5.4. (a) Waveguide chip inside the petri dish. (b) Schematic diagram of the cross-sectional view of laser light coupled and guided towards the specimen in optical waveguides. (c) Exciation and collection of path in chip based nanoacopy to perform dSTORM.

polished silicon wafer with a monolayer of silicon dioxide (SiO_2) followed by a thin layer of silicon nitride (Si_3N_4). Fabricating the PIC chip with side walls on the top surface works as an efficient waveguide structure for light propagation into the specimen. The objective lens is used to focus light onto the end-facet, and the laser light propagates inside the waveguide structure via total internal reflection phenomenon. A small amount of electric field of light leaks out which is known as an evanescent wave. Similar to the TIRF microscopy, intensity of the evanescent field can be expressed as:

$$I(z) = I_0 \exp(-z/d) \tag{5.4}$$

where I_0 refers to the incident intensity and $I(z)$ is the intensity of the evanescent field at distance z (perpendicular to the interface). d is the characteristic depth of penetration given by

$$d = \frac{\lambda}{4\pi\sqrt{n_1^2 \sin^2\theta - n_2^2}} \tag{5.5}$$

It can be seen from equation (5.4), that the depth of penetration evanescent field is determined by the wavelength of source and local refractive index at the waveguide-sample boundary. Fluorophores in the fluorescently labelled sample are excited by the evanescent field. The sample mounting protocols are the same as for the glass slide. The waveguide dimension is crucial to produce sufficiently strong photo-switching of fluorophores on top of its surface. Ideally, the evanescent field intensity must be 1–10 kW cm^{-2} for efficient photo-switching of the fluorophores [23].

The intensity of evanescent field highly depends on the refractive index of the material (Si_3N_4, $n = 2$ for 660 nm) [24]. Standard protocols to fabricate the waveguide chips can be found elsewhere [23]. A different mode that travels inside the waveguide travels discrete paths thus generating non-uniform patterns. These patterns can be averaged, and thus create a uniform image by shifting the mode pattern in time. Rapid scanning of the input beams using the galvo scanner can be used to achieve different mode patterns at different times inside the waveguide. In the case of dSTORM, thousands of images are required to localize each molecule in the sample and thus to achieve a super-resolved image. All these images can be reconstructed using ThunderSTORM [25], a FIJI plugin to generate the super-resolution image.

5.3.3 Chip preparation and staining protocol

Cleaning of a waveguide chip is a two-step process. The first step is to clean the chips with % Hellmanex in DI water at 70°C for 10 min After that, the chip is rinsed with DI water, followed by iso-propanol. The chip should be again cleaned by DI water and finally dried using N_2. A rectangular hollow chamber can be created with polydimethylsiloxane (PDMS) and placed on the chip to restrict the area to attach the cells. The attached cells can be stained with any fluorescent dye, for example cellmask deep red. In addition, a dSTORM buffer can be prepared using 22 μl PBS, 5 μl MEA, 2.5 μl enzymes and 20 μl glucose. The sample is then rinsed with PBS before mixing blinking buffer. Finally, the sample area can be sealed off with a cover slip for imaging.

5.4 Integrated on-chip nanoscopy and partially spatially coherent quantitative phase microscopy

A schematic diagram of integrated partially spatially coherent QPM and chip based nanoscopy set-up is shown in figure 5.5. Cobolt laser (@ 560 nm) is used in the experimental system to perform quantitative phase imaging (QPI). Source light is first expanded by a microscopic objective (MO_2) and then passes through the rotating diffuser. The diffused light after the diffuser is coupled into multimode fibre (MMFB) using lens L_3. The output beam of MMFB results in reduced spatial coherence of the light beam. The light from MMFB output is coupled into the input of the QPM system. The extended monochromatic light passes through a lenses L_4 and L_5 to focus it into the back focal plane of MO_3 and MO_4. The light beam reflected from the sample interferes with the reference beam at the beam splitter. The interferogram is finally projected onto the complementary metal oxide semiconductor image sensor using a tube lens. dSTORM imaging can be done using chip based TIRF excitation. The stained sample with blinking buffer can be placed on the sample stage with a vacuum chuck to avoid any mechanical vibration. The excitation light is, acoupled from free space using an objective lens (MO_1). Depending on the width and height of the waveguides it gives an inhomogeneous excitation pattern. Typically multimode waveguide is used to perform the imaging and the coupling objective is scanned along the input facet to average out the modes and to achieve homogeneous resolution. Initially, less laser power can be used to excite the blinking fluorophore which is further incrementally increased up to approximately 50 mW to get additional localizations. The data were reconstructed using ThunderSTORM with a FIJI plugin.

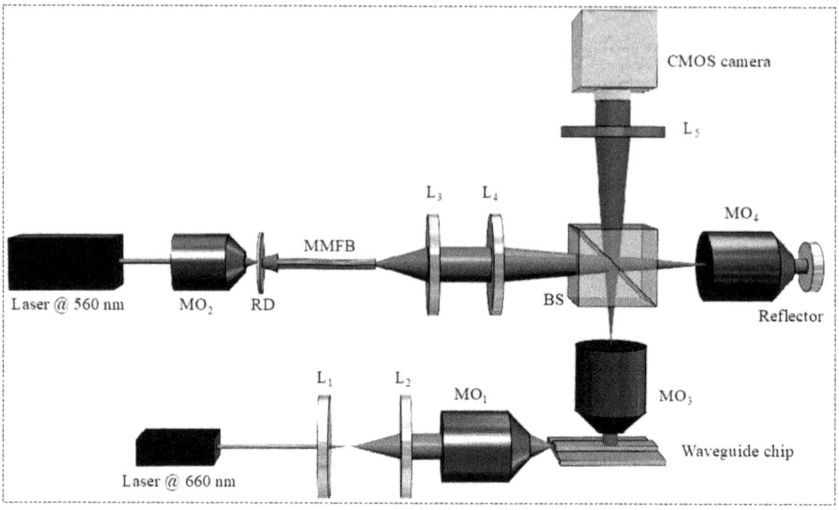

Figure 5.5. Schematic diagram of integrated partially spatially coherent quantitative phase microscopy and on-chip nanoscopy system. MO: objective lens, RD: rotating diffuser, BS: beam splitter, L: lens, MMFB: multi-multimode fibre bundle. Evanescent field is generated on top of the waveguide chip for single-molecule fluorescence excitation. An upright microscope is converted into a Linnik type interferometer to acquire both fluorescence imaging and perform QPI. Reprinted with permission from [26], copyright 2021 the Author(s). Published by PNAS.

5.5 Applications of integrated on-chip nanoscopy and quantitative phase microscopy

Figure 5.6 depicts the bright field, interferometric image, and the phase map of liver sinusoidal endothelial cells (LSECs). The bright field image (figure 5.6(a)) of the LSECs was recorded simply by blocking the reference arm of the interferometer. Since it represents the intensity value only, it is not possible to locate different parts of the cells, as can be seen in figure 5.6(a). The interferometric image of the same location is shown in figure 5.6(b). A phase map from the interferometric image is reconstructed using standard Fourier transform and transport of intensity algorithm. The reconstructed phase map of the LSECs is depicted in figure 5.6(c). The color map represents the phase value in radian, i.e., the optical thickness which is the combined information of RI and thickness of the sample. Deep red represents the maximum phase map and deep blue corresponds to background. A phase map of the nucleus part is found to be maximum (2–4 rad) and varies for the cell membrane between 100 and 200 mrad.

Figure 5.7 depicts a super-resolved image of LSECs reconstructed from the raw data acquired from the integrated system while operating in a nanoscopy mode. Figure 5.7(a) shows the bright field image of the LSECs. After following each step mentioned in dSTORM imaging and data analysis section, the super-resolution image of the highlighted black dotted region of LSECs is shown in figure 5.7(b). Fine features of the thinnest part, i.e., membrane is zoomed and shown in figure 5.7(c). The small holes in the cell membrane are known as fenestrations. These fenestrations

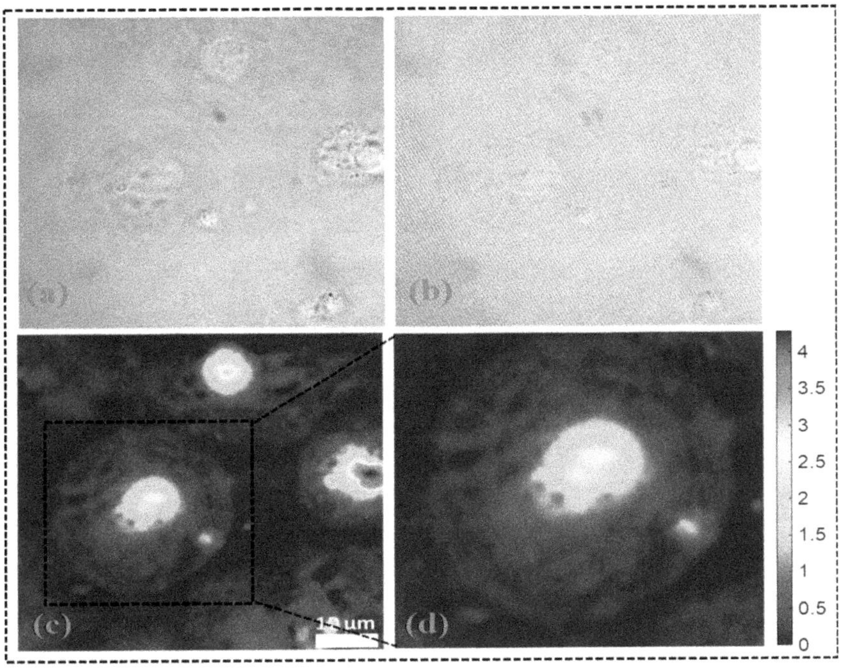

Figure 5.6. Experimentally acquired images of LSECs. (a) Bright field image by blocking the reference arm of the interferometer, (b) interferometric image of LSECs. The reconstructed phase (c) and the zoomed images (d) show phase maps of one LSEC. Axes of the images are in pixels and colorbar represents the phase value 'ϕ' in radian. Reproduced with permission from [26], copyright 2021 the Author(s). Published by PNAS.

transfer substrates (particularly insulin and lipoproteins) between blood and liver cells. They play an important role in aging and the regulation of cholesterol (lipoproteins). Size of the fenestrations typically varies between 50 and 150 nm and in general they are arranged in a group called sieve plate. It is important to note that the size of fenestrations is much smaller than the diffraction limit and thus they can be visible only by optical nanoscopy techniques. Electron microscopy can also be used to visualize these fenestrations but due to special sample preparation, cannot offer temporal information of the cells.

The use of super-resolution fluorescence microscopy to probe into several biological phenomena is rapidly expanding into cells, tissues, and molecular biology. Different super-resolution techniques such as total internal reflection dark-field (TIRD), confocal microscopy, SIM, STED, PALM, dSTORM have been proposed in recent years for several biological applications. Thus, integrating a label-free technique such as QPI with super-resolution microscopy can offer chemical and morphological details of the specimens. Recently Butola *et al* [26] demonstrated the application of multimodal microscopy on liver cells. In future, the system can be adapted for imaging live-cell dynamics. Being able to detect nanometric changes in the spatial direction using dSTORM and nanometric sensitivity in the axial direction using QPM, the current method will be useful in biology and biomedicine.

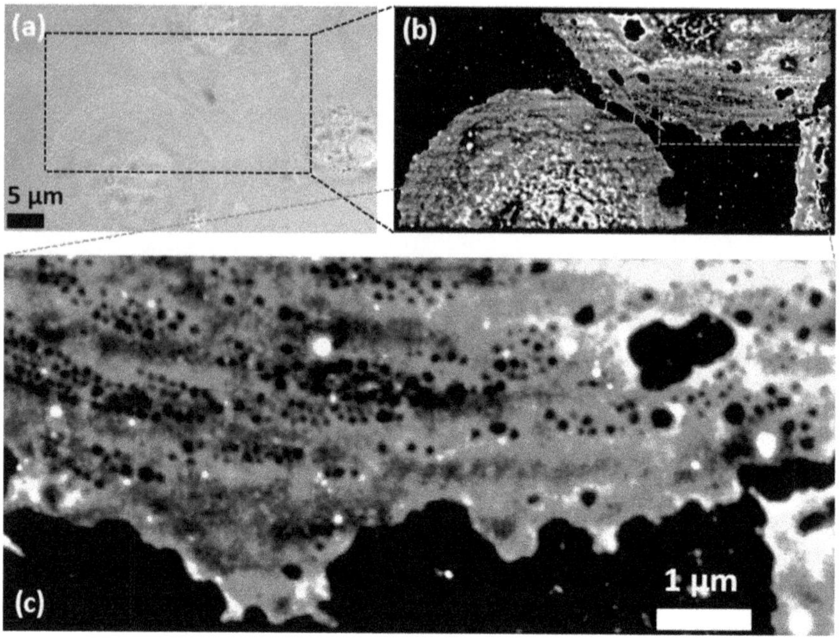

Figure 5.7. dSTORM image of LSECs reconstructed from the raw data acquired from integrated partially spatially coherent quantitative phase microscopy and chip based nanoscopy system. (a) Bright field image of LSECs, (b) dSTORM image of cropped black dotted box highlighted in the bright field image and the zoomed area (c) of dSTORM image. Small holes known as fenestrations in the membrane part of the LSEC can be seen here. Size of these fenestrations varies between 30 and 200 nm. Reproduced with permission from [26], copyright 2021 the Author(s). Published by PNAS..

References

[1] Lichtman J W and Conchello J-A 2005 Fluorescence microscopy *Nat. Methods* **2** 910–19

[2] Herman B 2020 *Fluorescence Microscopy* (New York: Garland Science)

[3] Ji N 2017 Adaptive optical fluorescence microscopy *Nat. Methods* **14** 374–80

[4] Kubitscheck U 2017 *Fluorescence Microscopy: From Principles to Biological Applications* (New York: Wiley)

[5] Webb D J and Brown C M 2012 Epi-fluorescence microscopy *Cell Imaging Techniques* (Berlin: Springer) pp 29–59

[6] Fish K N 2009 Total internal reflection fluorescence (TIRF) microscopy *Curr. Protoc. Cytom.* **50** 12.18.11–13

[7] Schermelleh L *et al* 2019 Super-resolution microscopy demystified *Nat. Cell Biol.* **21** 72–84

[8] Sanderson M J, Smith I, Parker I and Bootman M D 2014 Fluorescence microscopy *Cold Spring Harbor Protocols* **2014** 1042–65

[9] Saleh B E and Teich M C 2019 *Fundamentals Of Photonics* (New York: Wiley)

[10] Ghatak A 2005 *Optics* (New York: Tata McGraw-Hill Education)

[11] Hell S W and Wichmann J 1994 Breaking the diffraction resolution limit by stimulated emission: stimulated-emission-depletion fluorescence microscopy *Opt. Lett.* **19** 780–82

[12] Gustafsson M G 2000 Surpassing the lateral resolution limit by a factor of two using structured illumination microscopy *J. Microsc.* **198** 82–7

[13] Shroff H, Galbraith C G, Galbraith J A and Betzig E 2008 Live-cell photoactivated localization microscopy of nanoscale adhesion dynamics *Nat. Methods* **5** 417–23
[14] Dertinger T, Heilemann M, Vogel R, Sauer M and Weiss S 2010 Superresolution optical fluctuation imaging with organic dyes *Angew. Chem. Int. Ed.* **49** 9441–43
[15] Van de Linde S *et al* 2011 Direct stochastic optical reconstruction microscopy with standard fluorescent probes *Nat. Protoc.* **6** 991
[16] Jost A and Heintzmann R 2013 Superresolution multidimensional imaging with structured illumination microscopy *Annu. Rev. Mater. Res.* **43** 261–82
[17] Fürstenberg A and Heilemann M 2013 Single-molecule localization microscopy–near-molecular spatial resolution in light microscopy with photoswitchable fluorophores *Phys. Chem. Chem. Phys.* **15** 14919–30
[18] Fiolka R, Shao L, Rego E H, Davidson M W and Gustafsson M G 2012 Time-lapse two-color 3D imaging of live cells with doubled resolution using structured illumination *Proc. Natl Acad. Sci.* **109** 5311–15
[19] Benke A and Manley S 2011 Live-cell dSTORM of cellular DNA based on direct DNA labeling *Chem. Bio. Chem.* **13** 298–301
[20] Alonas E *et al* 2014 Combining single RNA sensitive probes with subdiffraction-limited and live-cell imaging enables the characterization of virus dynamics in cells *ACS Nano* **8** 302–15
[21] van de Linde S, Sauer M and Heilemann M 2008 Subdiffraction-resolution fluorescence imaging of proteins in the mitochondrial inner membrane with photoswitchable fluorophores *J. Struct. Biol.* **164** 250–54
[22] Diekmann R *et al* 2017 Chip-based wide field-of-view nanoscopy *Nat. Photonics* **11** 322
[23] Helle Ø I, Coucheron D A, Tinguely J-C, Øie C I and Ahluwalia B S 2019 Nanoscopy on-a-chip: super-resolution imaging on the millimeter scale *Opt. Express* **27** 6700–10
[24] Coucheron D A, Wadduwage D N, Murugan G S, So P T and Ahluwalia B S 2019 Chip-based resonance Raman spectroscopy using tantalum pentoxide waveguides *IEEE Photonics Technol. Lett.* **31** 1127–30
[25] Ovesný M, Křížek P, Borkovec J, Švindrych Z and Hagen G M 2014 ThunderSTORM: a comprehensive imageJ plug-in for PALM and STORM data analysis and super-resolution imaging *Bioinformatics* **30** 2389–90
[26] Butola A *et al* 2021 Multimodal on-chip nanoscopy and quantitative phase imaging reveals the nanoscale morphology of liver sinusoidal endothelial cells *Proc. Natl Acad. Sci.* **118** e2115323118

Chapter 6

Longitudinal spatial coherence gated tomography using partially spatially coherent monochromatic light

In most of the optical coherence tomographic (OCT) techniques a temporally low (broad-band) and spatially highly (point source) coherent light is used to obtain high axial resolution. Various types of OCT systems are described and experimental results of tomography of an ocular lens and a multilayer sample using swept-source OCT are discussed. However, all these OCT systems require a dispersion compensation optics and due to high spatial coherence speckle noise is generated. In this chapter we demonstrate that by means of using partially spatially coherent extended monochromatic light source both speckle-free tomography and profilometry can be done. The concept of longitudinal spatial coherence (LSC) is described by means of using an analogy between temporal and spatial coherence and experimental results of LSC gated tomography and topography are discussed.

6.1 Introduction to time-domain and frequency-domain optical coherence tomography

OCT is non-contact and non-invasive cross-sectional high-resolution imaging of biological tissues *in vivo.* At the heart of an OCT system is low coherence interferometry in which high-resolution optical sectioning can be done using low temporal coherence of a light source, this is because the axial resolution of an OCT system is inversely proportional to spectral bandwidth of the light source. The first OCT system was realized in time-domain in 1991 by Hwang *et al* [1, 2]. Since then, a variety of OCT systems have been realized in time-domain [2–6], frequency-domain, spectral-domain and swept-source OCT [3–9].

6.1.1 Time-domain OCT

Figure 6.1(a) shows the schematic diagram for time-domain (TD)-OCT. In this system light from a broad-band (temporally low coherence) source is made incident

doi:10.1088/978-0-7503-3987-2ch6

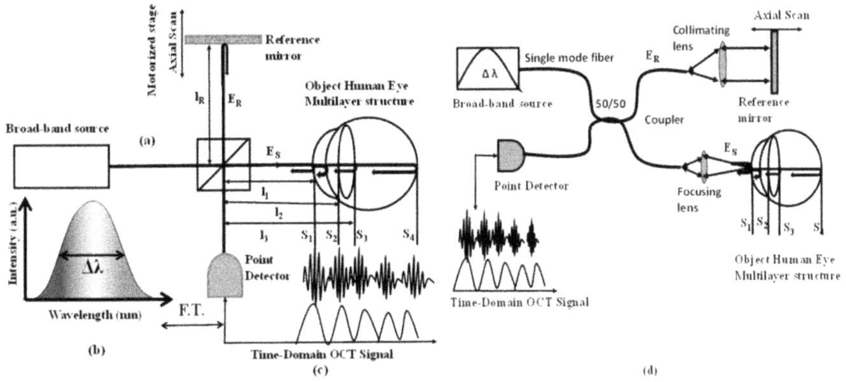

Figure 6.1. Schematic diagram illustrating TD-OCT system; (a) low-coherence interferometry system; (b) broad-band spectrum of light source and (c) time-domain OCT signal.

onto a 50:50 beam splitter (BS) and one beam travels towards the reference mirror (E_R) and another beam towards a multilayer sample (E_S) like a human eye. Light is back-reflected from each layer of the sample as shown in figure 6.1(a) [2–8]. The reference mirror is scanned in the axial direction and interference occurs between the reference and sample beams only when the optical path difference (OPD) between them is within the coherence length of the light source. Figure 6.1(b) shows the spectrum of the broad-band light source having spectral bandwidth, $\Delta\lambda$, large, therefore the coherence length (l_c) of such a source is very small. Due to this short coherence length the interference occurs between the reference beam and only one sample beam at a time for which the OPD < l_c, even though reflections from other layers are also present which serve as background. Due to this, selective occurrence of interference from one sample beam at a time leads to high signal-to-noise ratio and optical sectioning of multilayer sample that is non-contact and non-invasive. In other words the undesired reflections are rejected due to non-interference.

In TD-OCT, both ranging and imaging are achieved simultaneously. OCT measures the echo time-delay and the reflective response from different depths and is therefore capable of imaging below the top surface of the sample non-invasively. The ranging is done by means of scanning the reference mirror in axial direction and interferograms of the light backscattered from each layer. Temporal interferograms are recorded sequentially as shown in figure 6.1—I. From the peak positions of any two interferograms the gap/thickness of each layer can be computed first. This is ranging with low coherence interferometry. Figure 6.1(d) shows the fibre-optic version of the TD-OCT system which is most frequently used for OCT applications in the field. In this figure, broad-band light is coupled into a fibre-coupler and light is split into 50/50 ratio and propagates into the sample and reference arms separately. The reference beam $E_R(t)$ is reflected from a reference mirror, whereas the measurement or sample beam $E_S(t)$ is reflected from the biological specimen or tissue that is to be imaged [3–5]. The reflected light from the sample and reference arms are recombined at the coupler again and the interference signal is converted into an electrical signal by the detector.

The interference signal is then demodulated and converted from analog to digital form for computer signal and image processing. The reference mirror is scanned continuously to match path lengths of the reference beam and object beam from different layers of the sample. The detector records the intensity of the output beam, which is proportional to the square of the amplitude of electromagnetic field. If the distance that light travels in the reference arm is l_R and the corresponding path travelled in the sample arm is l_S, then the intensity of the recorded interference pattern at the detector can be expressed as [3].

If $E_r = \frac{E_i}{\sqrt{2}} r_{\text{ref}} e^{i2kz_r}$ and $E_s = \frac{E_i}{\sqrt{2}} \sum_{n=1}^{N} r_{Sn} e^{i2kz_{sn}}$ represent the electric field at the BS after reflecting from the reference and nth layer of the sample arm, respectively, the 1D interference signal at the detector can be written as [3–5]

$$I(k, \omega) = \frac{\mu}{2} \left\langle \left| \frac{s(k, \omega)}{\sqrt{2}} r_{\text{ref}} e^{i(2kz_r - \omega t)} + \frac{s(k, \omega)}{\sqrt{2}} \sum_{n=1}^{N} r_{Sn} e^{i(2kz_{sn} - \omega t)} \right|^2 \right\rangle \tag{6.1}$$

where $s(k,\omega)$ represents the electric field amplitude and μ is the responsivity of the detector (Amperes/Watt). r_{ref} and r_{Sn} shows the electric field reflectivity of the reference and nth layer of the sample arm, respectively. The above equation can be expanded as [3–5];

$$\begin{aligned} I(k, \omega) &= \frac{\mu}{4} [S(k)(R_{\text{ref}} + R_{s1} + R_{s2} + \ldots + R_{sn})] \\ &+ \frac{\mu}{2} \left[S(k) \sum_{n=1}^{N} \sqrt{R_{\text{ref}} R_{sn}} \left(e^{i2k(z_{\text{ref}} - z_{sn})} + e^{-i2k(z_{\text{ref}} - z_{sn})} \right) \right] \\ &+ \frac{\mu}{4} \left[S(k) \sum_{n \neq m=1}^{N} \sqrt{R_{sn} R_{sn}} \left(e^{i2k(z_{sm} - z_{sn})} + e^{-i2k(z_{sm} - z_{sn})} \right) \right] \end{aligned} \tag{6.2}$$

In equation (6.2) the first term is background or DC component, second term is cross-correlation (represents OCT signal) and he last term is auto-correlation which is arising from the self-interference between the reflections from multiple layers.

The above equation (6.2) can be written as [3–5];

$$\begin{aligned} I(k, \omega) = \frac{\mu}{4} [S(k)(R_{\text{ref}} + R_{s1} + R_{s2} + \ldots + R_{sn})] + \frac{\mu}{2} \\ \left[S(k) \sum_{n=1}^{N} \sqrt{R_{\text{ref}} R_{sn}} \left(2\cos(2k(z_{\text{ref}} - z_{sn}))\right) \right] + \frac{\mu}{4} \\ \left[S(k) \sum_{n \neq m=1}^{N} \sqrt{R_{sn} R_{sn}} \left(2\cos(2k(z_{sm} - z_{sn}))\right) \right] \end{aligned} \tag{6.3}$$

Here, $S(k)$ represents Gaussian spectral profile, i.e., square of the electric field amplitude encodes the spectral dependence of the light source. In addition, the degree of coherence for a light source with Gaussian spectral profile is given by [2–4];

$$\gamma(\tau) = \exp\left[-\left(\frac{\pi\Delta\nu\tau}{2\sqrt{ln2}}\right)^2\right] . \exp(-j2\pi\nu_0\tau) \tag{6.4}$$

where $\Delta\nu$ is the frequency width of the temporal frequency spectrum and ν_0 is the central frequency of the light source. The key point in an OCT imaging system is the dominance of coherence function over axial point-spread function (at least for low numerical aperture lens) hence the axial resolution of an OCT system is defined by the coherence length of the light source [6–9].

In TD-OCT, the reference mirror is translated continuously along the beam axis to acquire the depth information of the sample. The 1D interference signal from multiple depths can be written as [3–5]:

$$I(z_r) = \frac{\mu}{4}\left[S\left(R_{\text{ref}} + R_{s1} + R_{s2} + \ldots + R_{sn}\right)\right] + \frac{\mu}{2} \left[S\sum_{n=1}^{N}\sqrt{R_{\text{ref}}R_{sn}}\, e^{-(z_{\text{ref}} - z_{sn})^2\Delta k^2}\cos[2k_0(z_{\text{ref}} - z_{sn})]\right] \tag{6.5}$$

The sample reflectivity profile is convolved with the source function and modulated by the difference between sample and reference arm and the central wavenumber of the source. Additionally, S represents integrated spectral power of the source and can be written mathematically as [2–4]

$$S = \int_0^\infty S(k)\, dk \tag{6.6}$$

In TD-OCT continuous scanning of the reference mirror (called A-scan) and X–Y scanning of object beam across the area of the sample (called B-Scan) are also required. Therefore, the speed of the data acquisition is slow and number of sampling points is limited. Due to these shortcomings the TD-OCT has poorer spatial resolution and low signal-to-noise ratio (SNR). To overcome these limitations frequency-domain (spectral-domain version) OCT was developed.

6.1.2 Frequency-domain (spectral-domain) optical coherence tomography

In SD-OCT, in place of a point photodetector a spectrometer and linear charge coupled device (CCD) array detector is used. In SD-OCT, a spectral interferogram between the backscattered light and the reference light is recorded using a spectrometer and a high-speed line-scan camera [19]. Figure 6.2(a) shows a schematic diagram of an SD-OCT system. The broad-band light, figure 6.2(b), passes through a Michelson interferometer and the spectral interferogram is recorded by the spectrometer at the output. The spectral interferogram is shown in figure 6.2(c), depicting the nature of spectral modulations corresponding with reflections from various depths of the multilayer sample. The depth information is

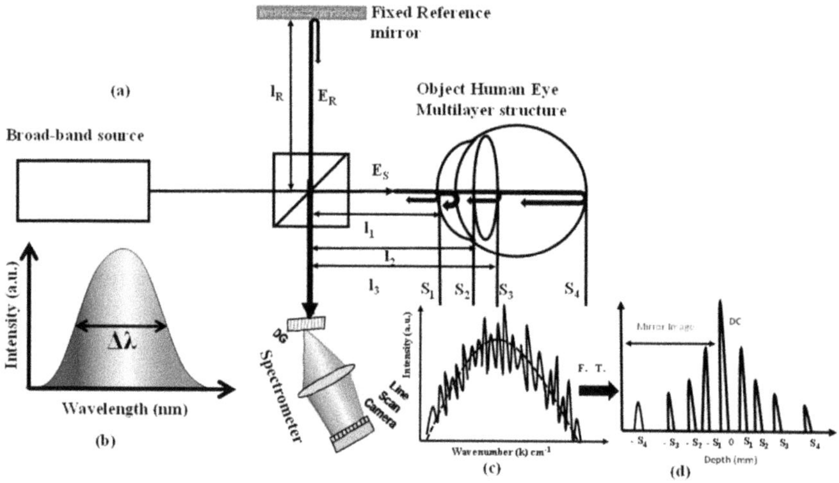

Figure 6.2. Schematic diagram illustrating frequency-domain (spectral-domain) OCT system; (a) low-coherence spectral interferometry system; (b) broad-band spectrum of light source, (c) spectral-domain OCT signal and (d) Fourier spectrum of spectral interferogram depth versus intensity.

acquired by analysing the backscattered signal reflected from different layers of the sample. Rescaling the signal from wavelength to wavenumber (k), and Fourier transform (FT) analyses is used to generate axial scans, figure 6.2(d). The multiple peaks are representing the reflectance from different depths and their mirror images. In SD-OCT, the reflectivity of the sample, i.e., $r_s(z_s)$, can be calculated by taking the inverse FT of equation (6.2), mathematically [3–10];

$$IFT[I(k,\omega)] = IFT\left[\frac{\mu}{4}\left[S(k)(R_{\mathrm{ref}} + R_{s1} + R_{s2} + \ldots R_{sn})\right] + \frac{\mu}{4}\left[S(k)\sum_{n=1}^{N}\sqrt{R_{\mathrm{ref}}R_{sn}}\left(e^{i2k(z_{\mathrm{ref}}-z_{sn})} + e^{-i2k(z_{\mathrm{ref}}-z_{sn})}\right)\right] + \frac{\mu}{4}\left[S(k)\sum_{n\neq m=1}^{N}\sqrt{R_{sn}R_{sn}}\left(e^{i2k(z_{sm}-z_{sn})} + e^{-i2k(z_{sm}-z_{sn})}\right)\right]\right] \tag{6.7}$$

Since cosine and delta function are Fourier transform pairs of each other, equation (6.7) can be simplified as [3, 4]

$$I(z) = \frac{\mu}{8}\left[\gamma(z)(R_{\mathrm{ref}} + R_{s1} + R_{s2} + \ldots R_{sn})\right] + \frac{\mu}{4}\left[\gamma(z)\otimes\sum_{n=1}^{N}\sqrt{R_{\mathrm{ref}}R_{sn}}\left(\delta(z \pm 2(z_{\mathrm{ref}} - z_{sn}))\right)\right] + \frac{\mu}{8}\left[\gamma(z)\otimes\sum_{n\neq m=1}^{N}\sqrt{R_{sn}R_{sn}}\left(\delta(z \pm 2(z_{sm} - z_{sn}))\right)\right] \tag{6.8}$$

where $[\gamma(z)(R_{\text{ref}} + R_{s1} + R_{s2} + \ldots R_{sn})]$ is the 'DC term', $\frac{\mu}{8}\left[\gamma(z) \otimes \sum_{n=1}^{N}\sqrt{R_{\text{ref}}R_{sn}}\,(\delta(z \pm 2(z_{\text{ref}} - z_{sn})))\right]$ represents the 'auto-correlation term', i.e., correlation between nth and mth layer of the sample. Additionally, $\frac{\mu}{4}\left[\gamma(z) \otimes \sum_{n=1}^{N}\sqrt{R_{\text{ref}}R_{sn}}\,(\delta(z \pm 2(z_{\text{ref}} - z_{sn})))\right]$ shows the 'cross-correlation term' where the desired sample field reflectivity, i.e., $\sqrt{R_s(z_s)}$ is correlated with the reference field. Using the sifting property of the delta function [2–5], i.e.

$$\gamma(z) \otimes \delta(z - z_0) = \gamma(z - z_0) \tag{6.9}$$

Equation (6.8) is referred to as 'A-scan' and can be written as [3, 4]

$$I(z) = \frac{\mu}{8}[\gamma(z)(R_{\text{ref}} + R_{s1} + R_{s2} + \ldots R_{sn})] + \frac{\mu}{4}\left[\sum_{n=1}^{N}\sqrt{R_{\text{ref}}R_{sn}}\,[\gamma[2(z_{\text{ref}} - z_{sn})] + \gamma[-2(z_{\text{ref}} - z_{sn})]\,]\right] + \frac{\mu}{8}\left[\sum_{n \neq m=1}^{N}\sqrt{R_{sn}R_{sn}}\,[\gamma[2(z_{sm} - z_{sn})] + \gamma[-2(z_{sm} - z_{sn})]\,]\right] \tag{6.10}$$

SD-OCT has very high sensitivity and speed advantage over TD-OCT because no depth-scan is required in this case.

6.1.3 Swept-source optical coherence tomography (SS-OCT)

The most advanced version of FD-OCT is swept-source SS-OCT system [3, 10, 11]. Figure 6.3(a) shows the schematic diagram of SS-OCT system. In this system light from a

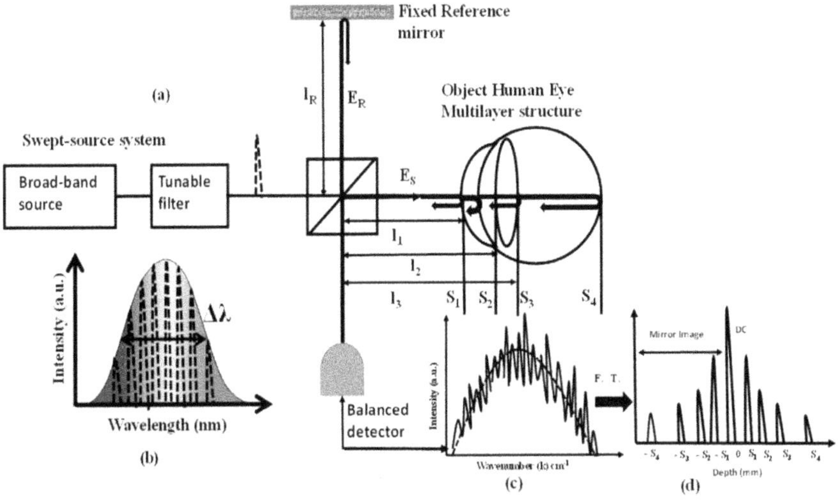

Figure 6.3. Schematic diagram representing SS-OCT system.

broad-band source (super-luminescent diode (SLD) or supercontinuum (SC) source) spectrum (figure 6.3(b)) passes through a fast spectral tunable filter, such as an acousto-optic tunable filter (AOTF) or a liquid-crystal tunable filter (LCTF) or a fast moving grating based mono-chromator. The input broad-band source with tunable filter used in the SS-OCT system is equal to the ouput spectrometer system of an SD-OCT system. The fine-tuned spectral components of broad band light are passed through the OCT engine, as shown in figure 6.3, and a balanced photodetector is used at the output of the OCT system. SS-OCT, also known as wavelength tuning interferometry (WTI), sequentially records the spectrally resolved intensity from different depth points of the sample. Frequency sweeping with respect to time interference and magnitude of the light signal defines the amplitude reflectance of light re-emitting sites in the sample. The sensitivity and speed of SS-OCT systems are mainly limited by the available sweeping tunable filter technology [10]. The main advantage of SS-OCT over the other OCT systems is the large depth of penetration, which is defined as [3, 10, 11];

$$z = \frac{\lambda^2}{4n\ \delta\lambda} \tag{6.11}$$

where λ is the peak wavelength of light source, $\delta\lambda$ represents instantaneous linewidth of the light source and n is the refractive index of the sample. In SS-OCT narrow bandwidth frequency of the broad-band spectrum is scanned as a function of time. The light backscattered from the multiple layers of the sample will have different frequency offsets compared to the light reflected from the reference mirror. When the backscattered light from the sample and the light reflected from the reference mirror interfere, the intensity modulation or beat intensity is generated at a frequency related to the frequency off-set. Due to this frequency, off-set different eco-time delays will generate different frequency modulations which are finally recoded by a balanced photodetector. Figure 6.3 I and (d) show the typical output of the SS-OCT system which is quite similar to SD-OCT system. From the spectral modulations the OCT images are reconstructed. The axial and transverse resolution of an SS-OCT system is similar to those of an SD-OCT system, however, the depth of penetration is better in the SS-OCT system given by equation (6.11) [3, 10, 11].

Figures 6.4(a) and (b) show the experimental results of ocular lens and Rat Eye, respectively which were recorded using MEMS-VCSEL Swept Source OCT (OCS1310V1—1300 nm, Thorlabs) system. Multilayer structures and shapes are clearly visible in both Ocular lens and Rat Eye. This kind of cross-sectional imaging of multiple internal layers with large depth and high resolution is possible while using SS-OCT. Figures 6.5(a) and (b) show the SS-OCT images of N95 multilayer face masks (a) Make 1 and (b) Make 2. A total of five layers of face masks are visible; therefore OCT can be used.

The SS-OCT system uses a unidirectional wavelength sweeping range from 1285 nm to 1315 nm with central wavelength of 1300 nm and an average output power incident on the sample was 25 mW. Source includes Mach–Zehnder interferometer 'k-clock' for optical clocking data acquisition. The sample arm of the SS-OCT system has an objective lens (5×, focal length ~25.1 mm in air) to image the specimen under study. Further, the interference signal was detected by a dual-balanced

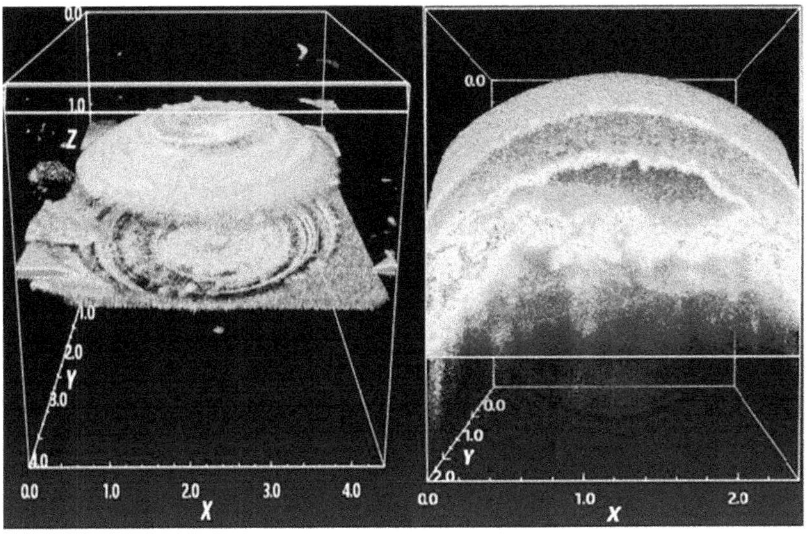

(a) (b)

Figure 6.4. SS-OCT images of ocular lens (a) and Rat Eye (b).

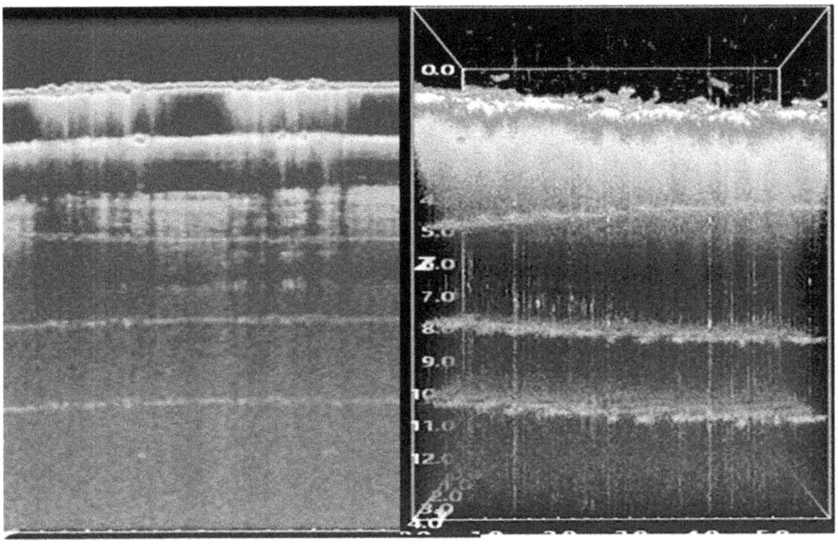

(a) (b)

Figure 6.5. SS-OCT images of N95 face masks (a) Make 1 (b) Make 2.

photodiode at 100 KHz A-Scan rate. By using two-dimensional XY scanners, one can scan the probe beam over the sample to obtain the 3D images, as shown in figures 6.4(a) and (b). The axial and transverse resolution of the SS-OCT system is 12 μm and 22 μm in air, respectively [12].

6.2 Axial-resolution in OCT systems

The axial resolution (Δz) of the OCT system is determined by the coherence length of the light source [1–6, 12].

$$\Delta z = \frac{lc}{2} = \frac{2ln2}{\pi}\frac{c}{\Delta \nu} = \frac{2ln2}{\pi}\frac{\lambda^2}{n(\lambda)\,\Delta\lambda} = 0.44\frac{\lambda^2}{n(\lambda)\Delta\lambda} \tag{6.12}$$

where λ is the central wavelength of the source. As can be seen from equation (6.12), the axial resolution of OCT system is inversely proportional to the spectral bandwidth of the light source. Therefore, the larger the spectral bandwidth of the source, the higher is the axial resolution of OCT. Thus the axial resolution of an OCT system is independent of the NA of the objective lens. On the other hand the lateral (transverse) resolution of an OCT is the same as for conventional optical microscopy and is determined by the NA of the lens. Transverse resolution is determined by the Gaussian beam waist at objective focus [1–6].

$$\Delta x = \left(\frac{4\lambda}{\pi}\right)\left(\frac{f}{D}\right) = \frac{2\lambda}{\pi}\frac{1}{NA} \tag{6.13}$$

where D is the spot size on the objective lens and f is its focal length. High transverse resolution can be obtained by using a large NA and focusing the beam to a small spot size. Depth of focus is confocal parameter [4]. In most of the OCT systems a low NA objective lens is used in the sample arms to achieve large depth of focus, figure 6.6. Therefore, the axial and lateral resolution of the imaging system is poor, however, in OCT systems high axial resolution is achieved due to short coherence length of the light source given by equation (6.12). Therefore, a thick sample can be optically sectioned by means of high axial resolution of the OCT system. As can be seen from

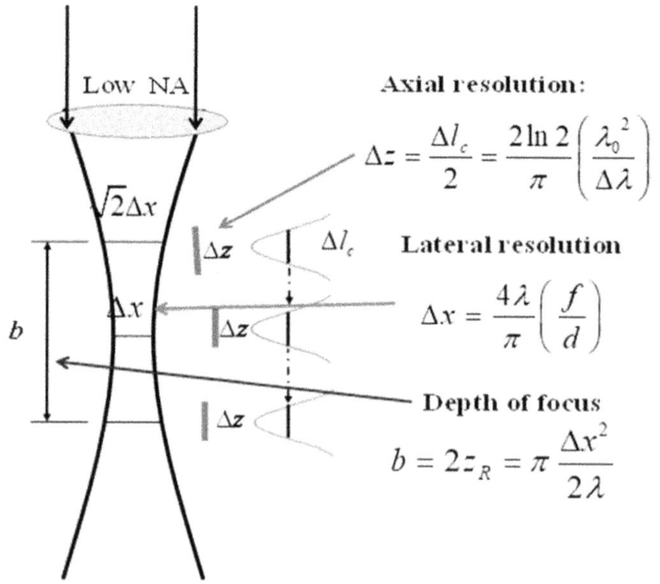

Figure 6.6. Depicting axial, and lateral resolution and depth of focus in OCT system.

figure 6.6, the thick sample can be sectioned into thin layers optically (non-invasively), thus inner structures of the multilayer thick samples can be imaged [3].

6.3 Dispersion effects in OCT

The use of broad-band illumination in a conventional OCT system gives high axial resolution but it also suffers from chromatic dispersion [2, 3, 13]. This requires the use of a dispersion compensation mechanism in all the types of OCT systems making them costly and complicated. Further, if the interspace between the sample and objective is not occupied with a suitable index matching material, the use of a high NA objective lens combined with broad-band illumination degrades the interference signal contrast in deep imaging zones. In addition, the sample through which the broad-band light propagates may have an inhomogeneous spectral response. Thus, either the light phase (due to dispersion) or amplitude (due to inhomogeneous absorption) might be changed. Due to the dispersion effects the axial resolution is refractive index dependent as can be seen from equation (6.12). The use of large spectral bandwidth of the light source leads to dispersion effect and refractive index of the sample is wavelength dependent and is given by expression [2, 3, 13]:

$$n(\lambda) = c\frac{dk}{d\omega} = n - \frac{dn}{d\lambda}. \tag{6.14}$$

From equations (6.12) and (6.14) it is clear that dispersion also affects the axial resolution of OCT. If the dispersion is zero then $dn/d\lambda$ is zero, which simply gives $n(\lambda) = n$, where n is the refractive index of the sample. To overcome this problem, full field (FF) approach with an area detector like a CCD camera is used for parallel or *en face* imaging of the sample. The high spatial coherence of the SLD light source as well the highly scattering surface of biological samples generates speckles which can degrade the quality of the image. In this chapter we describe the use of longitudinal spatial coherence of extended quasi-monochromatic light source to obtain high axial resolution OCT imaging without using a broad-band light source. To accomplish this a low cost laser with partial spatial coherence is required.

6.4 Concept of longitudinal spatial coherence

6.4.1 Analogy between temporal and spatial coherence

Temporal coherence, which is the basis behind profilometry and tomography, is associated with the term longitudinal temporal coherence. This is because the coherence between two points along the propagation axis is determined by the radiation's temporal spectrum, as is manifested by the Wiener–Khinchin theorem, which is depicted in figure 6.7 [2–5];

$$\gamma(\tau) = \int_{-\infty}^{\infty} S(\nu)\exp(i2\pi\nu\,\tau)df \tag{6.15}$$

where $\gamma(\tau)$ is the temporal coherence function of the source and $S(\nu)$ is the source spectral distribution function (power spectral density). However, according to the

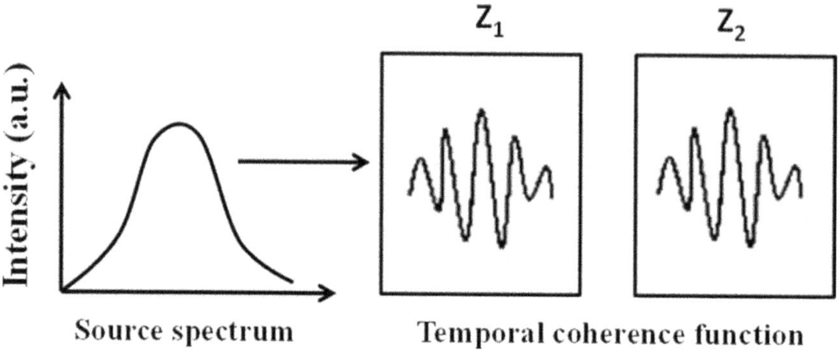

Figure 6.7. Illustrating the Wiener–Khinchin theorem for source spectrum and temporal coherence function.

van Cittert–Zernike theorem the longitudinal spatial coherence (LSC) function and the longitudinal component of the spatial frequency are FTs of each other. Thus the coherence between two points along the propagation axis can also be determined by changing the spatial (rather than temporal) structure of a quasi-monochromatic incoherent planar source [8, 9]. To realize the LSC the spatial coherence of the light source is engineered and by making it spatially extended and partially spatially coherent.

6.4.2 Longitudinal spatial coherence

If the light source is extended and broad-band then the light generated by such a source will have both low spatial and temporal coherence. Further, the spatial coherence can be two types, i.e., lateral spatial coherence (if the light vibrations arriving at two space points P_1 and P_2 at the same time in the lateral directions are in phase) and longitudinal spatial coherence (if the light vibrations arriving at two space points Z_1 and Z_2 the same time in the axial direction are in phase). Under these situations the generalized interference equation for this light can be written as [14–20];

$$I(x, y) = I_1(x, y) + I_2(x, y) + 2\sqrt{I_1(x, y) + I_2(x, y)}\ \gamma_{12}(\Delta r, \Delta z, \Delta t)\cos\left\{\frac{2\pi}{\lambda} 2\Delta z + \Delta\phi_{12}(r, \Delta z)\right\} \tag{6.16}$$

where $\gamma_{12}(\Delta r, \Delta z, \tau)$ is the normalized degree of spatial, temporal and longitudinal spatial coherence function.

For a quasi-monochromatic and spatially incoherent extended light source, the longitudinal spatial coherence is dominant and in this case the interference law equation (6.16) can be written as [14–20]; In this case the interference contrast is governed by LSC only.

$$I(x, y) = I_1(x, y) + I_2(x, y) + 2\sqrt{I_1(x, y) + I_2(x, y)}\ \gamma_{12}(\Delta r = 0, \Delta z, \Delta t = 0)\cos\left\{\frac{2\pi}{\lambda} 2\Delta z + \Delta\phi_{12}(r, \Delta z)\right\} \tag{6.17}$$

The generalized form of van Cittert–Zernike theorem relates longitudinal spatial coherence (LSC) function to the spatial structure (i.e., angular frequency spectrum) of the quasi-monochromatic extended light source. Analogous to the Wiener–Khintchine theorem [20–24], the generalized van Cittert–Zernike theorem states that LSC function '$\Gamma(\delta z, \Delta t = 0)$' and source angular frequency spectrum form Fourier transform pairs. According to this theorem, correlation or LSC function '$\Gamma(\delta z, \Delta t = 0)$' is defined as follows [20–24]:

$$\Gamma(\delta z, \Delta t = 0) = \int_{-\infty}^{\infty} S(k_z)\exp(ik_z\delta z)dk_z, \tag{6.18}$$

where $\Gamma(\delta z, \Delta t=0)$ is the longitudinal spatial coherence function and $S(k_z)$ is the angular frequency spectrum of the light source, δz ($=z_1 - z_2$) is the separation between spatial points $Q_1(z_1)$ and $Q_2(z_2)$ situated in two different observation planes, and k_z is the longitudinal spatial frequency [20–24].

$$K_z = \frac{2\pi}{\Lambda_z} = \frac{2\pi}{\lambda}\cos\theta_z = \frac{\omega}{c}\cos\theta_z \tag{6.19}$$

where Λ_z is the spatial period along the z-direction, θ_z is the angle between the direction of propagating field and z-axis and λ is the wavelength of the light source. The longitudinal coherence length is defined as [20–24]:

$$L_c = \frac{2\pi}{\Delta k_z} \tag{6.20}$$

where Δk_z is the longitudinal spatial frequency range.

The general expression of the longitudinal coherence length (L_c) which depends on both the angular frequency and temporal frequency spectrum of the light source, as follows [20–24]:

$$L_c = \left[\frac{2\sin^2(\theta_z/2)}{\lambda} + \frac{\Delta\lambda}{\lambda^2}\cos^2(\theta_z/2)\right]^{-1} \tag{6.21}$$

where, θ_z is half of the angular spectrum width, λ is the central wavelength, and $\Delta\lambda$ is related to the temporal spectrum width of the source.

If the size of the light source is small (i.e. a point source) then the coherence length or the coherence time or simply the longitudinal coherence properties of the light vibrations can be purely determined by the temporal frequency spectrum (spectral distribution) of the light source [20–24]. On the other hand, if the size of the light source is large (extended) and temporal frequency spectrum is narrow (quasi-monochromatic, ($\Delta\nu \ll \nu_0$)) then the longitudinal spatial coherence properties of the light source can be purely determined by the angular frequency spectrum of the light source [20–24]. Thus, for sufficiently narrow temporal frequency spectrum width, the second term in the above expression can be neglected and L_c is determined by the angular frequency spectrum only:

$$L_c = \lambda/2\sin^2\left(\frac{\theta_z}{2}\right) \tag{6.22}$$

In this case, the axial resolution (i.e.,$L_c/2$) is dominated by LSC rather than temporal coherence length of the source. In the following section, this phenomenon is experimentally demonstrated and applied for QPM and high-resolution spatial coherence gated optical sectioning of industrial and biological cells.

6.5 Longitudinal spatial coherence gated topography and tomography.

The LSC properties of an extended quasi-monochromatic light source can be utilized for coherence gated tomography and topography of industrial and biological samples, even though the temporal coherence of light is large. Under the special conditions the occurrence of high contrast interference fringes is only due to LSC of the source rather than the temporal coherence [20–33]. The experimental set-up to realize LSC is shown in figure 6.8. In this set-up, light from a diode pumped solid state (DPSS) laser passes through an MO lens MO1 and the expanded beam illuminates a static ground glass diffuser. The scattered light is collected by a lens and coupled to a 2-meter-long multimode fibre bundle (MMFB) containing thousands of fibres in each multimode. The MMFB is vibrated to scrabble the modes further. The size of the MMFB is 12 mm and the output speckle-free light is collimated by lens L1 and L2 and the spatially incoherent light is then made incident onto the BS. The two beams are then passed through a Linnik-type interferometer containing MO_2 and MO_3. MO_2 and reference mirror are mounted on a linear motorized stage for scanning the reference mirror in the axial direction and MO_3 is used for focusing light onto the sample. The focus position of the sample can also be changed. The light reflected from the reference mirror and sample is made to interfere at the BS and the same is recorded by the CCD camera, images stored in the computer for further processing. First of all, we

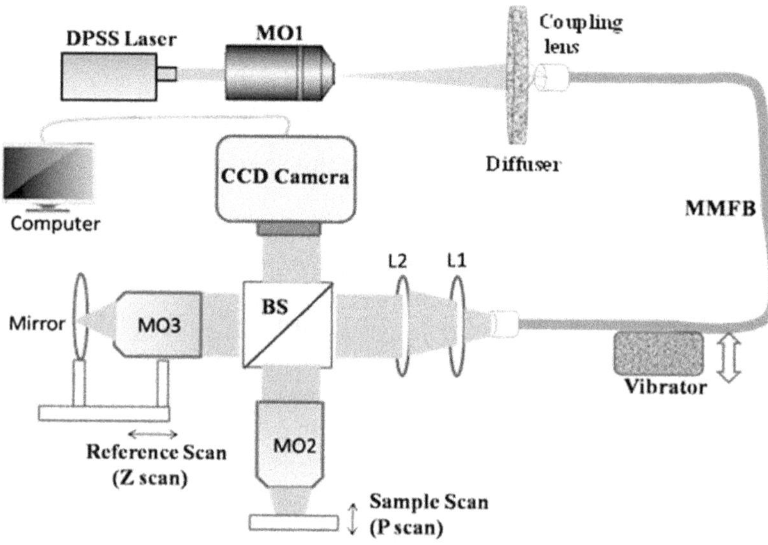

Figure 6.8. Schematic diagram of experimental set-up for LSC gated topography and tomography.

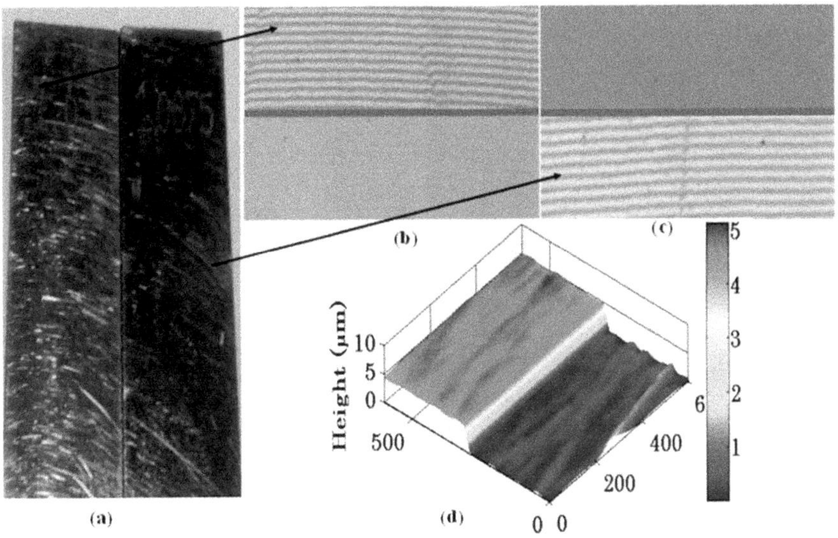

Figure 6.9. (a) Photograph of two gauge blocks with standard height, (b) high contrast interferogram on the left side of gauge block, (c) interferogram on the right side and (d) reconstructed heigh (topography) of the gauge block (colour bar represents height in micron).

conducted an experiment with standard height gauge blocks. Figure 6.9(a) shows the photograph of two gauge blocks with standard height. The height difference between the two gauge blocks was 5 μm.

The reference mirror was scanned in the axial direction and the high contrast interferogram occurred on the left side of the gauge block, figure 6.9(b), and no interferogram on the right side. We further scanned the reference mirror in the axial direction and the interferogram disappeared from the left side and a high contrast interferogram appeared on the right-side figure 6.9(c). The height gap is only 5 μm and the round-trip optical path difference is 10 μm. Although the temporal coherence length of the DPSS laser is 10 mm, we could optically section 5 μm step-height of the gauge block. This is due to the fact that LSC is dominant for an extended light source despite the fact that temporal coherence length is quite long. Therefore, the LSC of a quasi-monochromatic light source can be engineered to perform high axial resolution topography and tomography in addition to speckle-free imaging. The concept of LSC based optical sectioning was also utilized for multilayer biological samples. A multilayer sample was prepared as shown in figure 6.10(a). On the mirror, a layer of onion peel was fixed and on top of that a cover slip. The incident light will be reflected from each layer; top and bottom of cover slip, from the onion layer and finally from the mirror. The thickness of each layer is give in figure 6.10(b)—I. Now if we use long coherence length light source (coherence length in 10 mm range of more) and the interferometer system works only on the temporal coherence of the light source, then light reflected from each of these layers will interfere with the reference light, therefore, we will not be able to optically section this surface. However, figure 6.10(b)—I shows that as we scan the reference mirror path in the axial direction

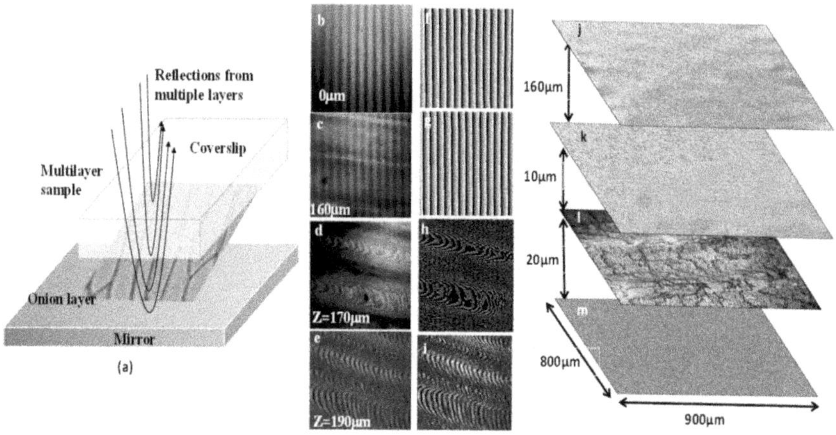

Figure 6.10. (a) Schematic of multilayer sample, (b)—I interferograms from each layer from top to bottom, respectively, (f)–(i) wrapped phase map corresponding to each interferogram, and (j)–(m) reconstructed optical sectioned images of multilayer sample.

the interferogram is obtained individually for each layer one by one. Therefore, we can use LSC of the light source for optical sectioning of the thin multiple layers. In addition to this, it can be seen from the interferograms that all of them are speckle-free with uniform illumination and uniform fringe contrast. Therefore, while using LSC one can perform high axial resolution topography and tomography of industrial as well as biological samples. Figures 6.10(j)–(m) are reconstructed optically sectioned images of a multilayer sample [20–23].

A reconstruction phase map in figures 6.9 and 6.10 was done using single-shot FT fringe analysis technique [34]. The two-dimensional interferogram can be expressed as;

$$I(x, y) = I_1(x, y) + I_2(x, y) + 2\sqrt{I_1(x, y)I_2(x, y)}\ |\gamma_{12}|\cos\left[\Delta\phi(x, y) + 2\pi f_0 x\right] \quad (6.23)$$

If I_1(x, y) = I_2(x, y) = I_0(x, y) then equation (6.23) can be written as

$$I(x, y) = b(x, y) + c(x, y)\left[\cos(\Delta\phi(x, y)) + 2\pi f_0 x\right] \quad (6.24)$$

where, $b(x,y)$ is the background (DC component), and $c(x,y)$ is the fringe modulation.

For simplicity, equation (6.24) can be written as

$$I(x, y) = b(x, y) + d(x, y)\exp(i2\pi f_0 x) + d^*(x, y)\exp(-i2\pi f_0 x) \quad (6.25)$$

where,

$$d(x, y) = \frac{c(x, y)}{2}\exp[i\Delta\phi(x, y)] \quad (6.26)$$

And

$$d^*(x, y) = \frac{c^*(x, y)}{2}\exp[i\Delta\phi(x, y)] \quad (6.27)$$

The FT of equation (6.25) is given by [34]

$$FT\left[I(x, y)\right] = B(f_x, y) + D(f_x - f_0, y) + D^*(f_x + f_0, y) \tag{6.28}$$

where f_0 is the carrier frequency. We select either D or D^* using the window and shifting it towards the origin, we can isolate the term which contains the desired information. By transforming the term $D(f_x, y)$ back into the original spatial domain using inverse FT we obtain $d(x,y)$. The real and imaginary phase are given by,

$$Re\{d(x, y)\} = d(x, y)\cos(\phi(x, y)) \tag{6.29}$$

$$Im\{d(x, y)\} = d(x, y)\cos(\phi(x, y)) \tag{6.30}$$

And the phase information can be expressed as

$$\phi(x, y, k) = \tan^{-1}\frac{Im\{d(x, y, k)\}}{Re\{d(x, y, k)\}} \tag{6.31}$$

Due to the use of the arctangent function, the extracted phase will be in the range $-\pi$ to π, and will need to be unwrapped using phase unwrapping algorithms. Phase-shifting fringe analysis can also be used for reconstructing the phase and amplitude images [20–33]. In summary, OCT can be performed using a spatially partially coherent monochromatic light source. There are two advantages of this over the conventional OCT system, i.e., no dispersion compensation optics is required and it is speckle-free imaging technology.

References

[1] Huang D, Swanson E A, Lin C P, Schuman J S, Stinson W G, Chang W, Hee M R, Flotte T, Gregory K and Puliafito C A 1991 Optical coherence tomography *Science* **254** 1178–81

[2] Bouma B E and Jearney G J 2002 *Handbook of Optical Coherence Tomography* (New York: Marcel Dekker, Inc.)

[3] Drexler W and Fujimoto J G 2008 *Optical Coherence Tomography Technology and Applications* 1st edn (New York: Springer)

[4] Fercher A F 1996 Optical coherence tomography *J. Biomed. Opt.* **1** 157–74

[5] Schmitt J M 1999 Optical coherence tomography (OCT): a review *IEEE J. Sel. Top. Quantum Electron.* **5** 1205–15

[6] Leitgeb R, Hitzenberger C and Fercher A F 2003 Performance of Fourier domain vs. time domain optical coherence tomography *Opt. Express* **11** 889–94

[7] Choma M A, Sarunic M V, Yang C and Izatt J A 2003 Sensitivity advantage of swept source and Fourier domain optical coherence tomography *Opt. Express* **11** 2183–9

[8] Fercher A F, Drexler W, Hitzenberger C K and Lasser T 2003 Optical coherence tomography-principles and applications *Rep. Prog. Phys.* **66** 239

[9] Leitgeb R, Drexler W, Unterhuber A, Hermann B, Bajraszewski T, Le T, Stingl A and Fercher A 2004 Ultrahigh resolution Fourier domain optical coherence tomography *Opt. Express* **12** 2156–65

[10] Srinivasan V J, Huber R, Gorczynska I and Fujimoto J G 2007 High-speed, high-resolution optical coherence tomography retinal imaging with a frequency-swept laser at 850 nm *Opt. Lett.* **32** 361–63

[11] Lim H, de Boer J F, Park B H, Lee E C W, Yelin R and Yun S H 2006 Optical frequency domain imaging with a rapidly swept laser in the 815–870 nm range *Opt. Express* **14** 5937–44

[12] Butola A, Joshi T, Ahmad A, Dubey V, Senthilkumaran P and Mehta D S 2018 3D topography and tomography of multilayered freeform optical surfaces using large-range measurement swept-source low-coherence interferometry *Laser Phys.* **28** 116101

[13] Wojtkowski M, Srinivasan V J, Ko T H, Fujimoto J G, Kowalczyk A and Duker J S 2004 Ultrahigh-resolution, high-speed, Fourier domain optical coherence tomography and methods for dispersion compensation *Opt. Express* **12** 2404–22

[14] Rosen J and Takeda M 2000 Longitudinal spatial coherence applied for surface profilometry *Appl. Opt.* **39** 4107–11

[15] Ryabukho V and Lyakin D 2005 The effects of longitudinal spatial coherence of light in interference experiments *Opt. Spectrosc.* **98** 273–83

[16] Ryabukho V, Lyakin D and Lobachev M 2005 Longitudinal pure spatial coherence of a light field with wide frequency and angular spectra *Opt. Lett.* **30** 224–6

[17] Ryabukho V, Lyakin D, Grebenyuk A and Klykov S 2013 Wiener–Khintchin theorem for spatial coherence of optical wave field *J. Opt.* **15** 025405

[18] McCutchen C 1966 Generalized source and the Van Cittert–Zernike theorem: a study of the spatial coherence required for interferometry *JOSA* **56** 727–33

[19] Abdulhalim I 2006 Competence between spatial and temporal coherence in full field optical coherence tomography and interference microscopy *J. Opt.* A **8** 952

[20] Mehta D S, Srivastava V, Nandy S, Ahmad A and Dubey V 2016 Full-field optical coherence tomography and microscopy using spatially incoherent monochromatic light *Handbook of Full-Field Optical Coherence Microscopy* (Singapore: Pan Stanford Publishing Pte. Ltd) ch 10 pp 357–92

[21] Ahmad A, Srivastava V, Dubey V and Mehta D 2015 Ultra-short longitudinal spatial coherence length of laser light with the combined effect of spatial, angular, and temporal diversity *Appl. Phys. Lett.* **106** 093701

[22] Dubois A 2016 *Handbook of Full-Field Optical Coherence Microscopy: Technology and Applications* (Boca Raton, FL: CRC Press)

[23] Ahmad A and Mehta D S 2018 Quantitative phase microscopy and tomography with spatially incoherent light *Advances in Optics: Reviews* ed S Yurish (Barcelona: International Frequency Sensor Association)

[24] Pavliček P, Halouzka M, Duan Z and Takeda M 2009 Spatial coherence profilometry on tilted surfaces *Appl. Opt.* **48** H40–47

[25] Safrani A and Abdulhalim I 2012 Ultrahigh-resolution full-field optical coherence tomography using spatial coherence gating and quasi-monochromatic illumination *Opt. Lett.* **37** 458–60

[26] Dubey S K, Anna T, Shakher C and Mehta D S 2007 Fingerprint detection using full-field swept-source optical coherence tomography *Appl. Phys. Lett.* **91** 181106–81108

[27] Dubey S K, Mehta D S, Anand A and Shakher C 2008 Simultaneous topography and tomography of latent fingerprints using full-field swept-source optical coherence tomography *J. Opt. A: Pure Appl. Opt.* **10** 015307–15

[28] Anna T, Shakher C and Mehta D S 2009 Simultaneous tomography and topography of silicon integrated circuits using full-field swept-source optical coherence tomography *J. Opt. A: Pure Appl. Opt.* **11** 045501–45509

[29] Anna T, Shakher C and Mehta D S 2010 Three-dimensional shape measurement of micro lens arrays using full-field swept-source optical coherence tomography *Opt. Lasers Eng.* **48** 1145–51

[30] Anna T, Srivastava V, Shakher C and Mehta D S 2011 Transmission mode full-field swept-source optical coherence tomography for simultaneous amplitude and quantitative phase imaging of transparent objects *IEEE Photon. Lett.* **23** 899–901

[31] Anna T, Srivastava V, Mehta D S and Shakher C 2011 High resolution full-field optical coherence microscopy using Mirau interferometer for the quantitative imaging of biological cells *Appl. Opt.* **50** 6343–51

[32] Subhash H M 2012 Full-field and single-shot full-field optical coherence tomography: a novel technique for biomedical imaging applications *Adv. Opt. Technol.* **2012** 435408

[33] Mehta D S, Naik D N, Singh R K and Takeda M 2012 Laser speckle reduction by multimode optical fiber bundle with combined temporal, spatial, and angular diversity *Appl. Opt.* **51** 1894–904

[34] Takeda M, Ina H and Kobayashi S J J 1982 Fourier-transform method of fringe-pattern analysis for computer-based topography and interferometry *J. Opt. Soc. Am.* **72** 156–60

IOP Publishing

Quantitative Phase Microscopy and Tomography
Techniques using partially spatially coherent monochromatic light
Dalip Singh Mehta, Ankit Butola and Veena Singh

Chapter 7

Low-coherence (white light) interference microscopy with colour fringe analysis

White light interferometry (WLI) along with phase-shifting technology is widely used in optical profilometry of industrial objects, coherence radar and in optical coherence tomography (OCT). Recently it has also been used for quantitative phase microscopy of biological cells. There are two main advantages of using WLI in quantitative phase microscopy (QPM), i.e., it is speckle-free, and one can obtain three colour interferograms from a single white light interferogram. In this chapter we first explain phase-shifting WLI for red blood cells, and an algorithm for colour fringe analysis to reconstruct the phase map of biological cells for three different wavelengths, i.e., red, green and blue colour, while performing the experiment only once. Experimental results for healthy red blood cells, anaemic cells, cancer cells and *Escherichia coli* bacteria are presented.

7.1 Introduction to low coherence interferometry

Low coherence interferometry is basically interference of light with a broad-band (white light) source, i.e., temporally low coherence light [1–3]. Since broad-band light consists of multiple wavelengths (polychromatic), all the light vibrations are in phase for an extremely short period of time (femto-second range) and their coherence length is also $\leqslant 1$ μm. In the case of broad-band light sources, high contrast interference occurs only for a short optical path difference (OPD). Due to its short coherence length it has been used for eye length measurement, intraocular distances and as coherence radar or spectral radar and OCT [2–8]. Figure 7.1 shows a schematic diagram of a Michelson interferometer with broad-band light source. While using broad-band light source the interference occurs only when the following condition is satisfied [1–3];

$$\Delta t \times \Delta \nu = 1 \tag{7.1}$$

where $\Delta t = t_2 - t_1$, the time taken by light vibrations to travel a round-trip optical path L_r and L_s, respectively, and $\Delta \nu$ is full-width half-maximum of the light source

doi:10.1088/978-0-7503-3987-2ch7

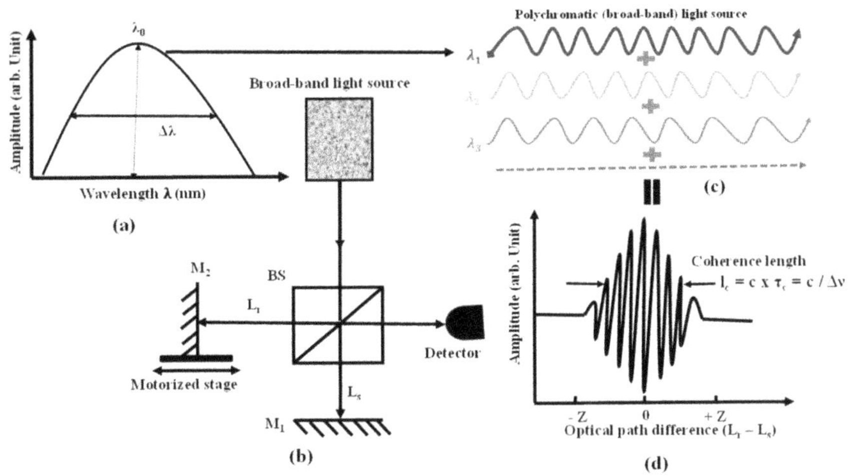

Figure 7.1. Schematic diagram depicting (a) broad-band light spectrum, (b) low coherence interferometric system using Michelson-type, BS: beam splitter, M_1 and M_2: mirrors, L_r and L_s are optical path lengths of two arms. (c) Polychromatic (broad-band) light with different wavelengths, and (d) the corresponding interferogram and coherence length.

spectrum. The time difference Δt must be less than equal to coherence time τ_c of the light source. The coherence time τ_c can be defined as the time duration within which the light vibrations are perfectly sinusoidal, or when two light vibrations arriving at the same spatial position P (at BS in figure 7.1) at two different times t_1 and t_2 are in phase or have constant phase difference then they interfere constructively. Using equation (7.1), $\tau_c = 1/\Delta\nu$, and coherence length $l_c = c \times \tau_c$. The coherence length of the light source can be defined as the path length within which the light vibrations are perfectly sinusoidal, or two light vibrations arriving at the same spatial position P (at BS in figure 7.1) at two different paths l_1 and l_2 are in phase or have constant phase difference then they interfere constructively. The coherence length l_c can be further expanded as follows [1–3];

$$l_c = c \; x \; \tau_c = c \,/\, \Delta\nu = \lambda^2/\,\Delta\lambda; \; (\text{using}\; \nu = c\,/\lambda; \; \Delta\nu = cx\; \Delta\lambda\,/\,\lambda^2) \qquad (7.2)$$

In general, the spectral profile of light sources is Gaussian, therefore, equation (7.2) can be rewritten as [1–3]:

$$l_c = \frac{2\ln 2}{\pi} \frac{\lambda^2}{\Delta\lambda} \qquad (7.3)$$

The coherence length of the light source is inversely proportional to the FWHM ($\Delta\lambda$) of the spectrum.

Therefore, the larger the $\Delta\lambda$, the smaller is the coherence length of the light source. For low coherence interferometry the coherence length is typically less than 100 μm. For example, for white light source with $\Delta\lambda = 250$ nm and peak wavelength $\lambda_0 = 550$ nm, the coherence length is less than 1.2 μm. Low coherence interferometry is widely used in OCT to obtain high axial-resolution optical sectioning of multilayer tissue structures [2, 3]. In OCT mainly broad-band light sources, such as white light and super-luminescent diodes (SLDs) are used, therefore, axial resolution in range 2–10 μm range can be obtained. In addition, low coherence interferometry can also be used for ranging and thickness measurement of multilayer samples in material science. White light interferometric profilers are most commonly used for 3D-surface profilometry for step-height measurement, and 3D-topography of industrial samples and optical components. The working principle of this equipment is based on low coherence interferometry and known as coherence scanning profilometry [9, 10]. Coherence scanning profilometry does not suffer any phase ambiguity problem, no spurious fringes are obtained, and it is speckle-free imaging. On the other hand, optical interferometric profilers based on coherent light suffer from a phase ambiguity problem, and are full of speckles and spurious fringes. Therefore, for measurement and imaging of all the realistic objects, industrial, biological or material science the most suitable technique is white light phase-shifting interferometry, by the use of which one can determine refractive index, optical path length change and many more biophysical parameters [11–30].

7.2 Phase-shifting white light interference microscopy

White light interference microscopy can be realised by means of using a bright-field microscope attached with a Michelson-type interferometric objective lens or Mirau-type interferometric objective lens. Both are dual beam interferometric objectives within build beam splitters and reference mirrors. The Michelson-type objectives are long working distance and low magnification (5× or 10×), while Mirau-type objectives are both long or short working distance objectives with magnifications 10×, 20×, 50× and 100×. These objectives can be easily mounted onto a bright-field microscope and an adapter, and a piezo-electric transducer (PZT) phase-shifter can also be attached to them, thus making a phase-shifting white light interference microscope. Since white light is composed of a minimum of three colors, red (R), green (G) and blue (B), therefore, superposition of these three colors leads to a colored white light interferogram. A white light interferogram recorded by color charge coupled device (CCD) camera can be expressed as [9–12];

$$\begin{aligned} I(x, y, \lambda_{R,G,B}) = {} & I_{obj}(x, y, \lambda_{R,G,B}) + I_r(x, y, \lambda_{R,G,B}) \\ & + 2\sqrt{I_{obj}(x, y, \lambda_{R,G,B}) I_r(x, y, \lambda_{R,G,B})} \\ & V\gamma(z) \cos [\phi(x, y, \lambda_{R,G,B}) + \delta(t)] \end{aligned} \tag{7.4}$$

where $\lambda_{R,G,B}$ are the peak wavelengths for red, green, and blue colour light interferograms in WLI, which can be digitally decomposed into individual interferograms, $I_{obj}(x, y, \lambda_{R,G,B})$, and $I_r(x, y, \lambda_{R,G,B})$ are the intensity of light from object and reference beams, V is the fringe contrast, $\gamma(\tau)$ is the temporal coherence of white light. In equation (7.4) $\phi(x, y, \lambda_{R,G,B})$ represents a 2D phase map and $\delta(t)$ is the controlled phase shift added purposely between reference and sample field to generate four-step phase-shifted interferograms. In general $I_{obj}(x, y, \lambda_{R,G,B}) = I_r(x, y, \lambda_{R,G,B}) = I_0(x, y, \lambda_{R,G,B})$, because the intensity of reference beam can be controlled and equalized with the intensity of object beam using a variable neutral density filter. Therefore, equation (7.4) can be rewritten in simplified form as follows [9–12];

$$I\ (x,\ y,\ \lambda_{\mathrm{R,G,B}}) = I_0(x,\ y,\ \lambda_{\mathrm{R,G,B}})\ [1 + V\gamma(z)\cos\ \{\phi(x,\ y,\ \lambda_{\mathrm{R,G,B}}) + \delta(t)\}] \quad (7.5)$$

To extract the 2D phase map $\phi(x, y, \lambda_{\mathrm{R,G,B}})$ of the object, five-phase-shifted interferograms, with a controlled phase shift $\delta_j(t)$ between reference and sample field is needed [9–12].π

$$\delta(\mathrm{t}) = -\pi,\ -\frac{\pi}{2},\ 0,\ \frac{\pi}{2},\ \pi; \quad (7.6)$$

For five-step phase-shifted interferograms using equations (7.5) and (7.6) they can be written as follows:

$$I_1(x,\ y,\ \lambda_{\mathrm{R,G,B}}) = I_0(x,\ y,\ \lambda_{\mathrm{R,G,B}})[1 + V\gamma(z)\cos\{\varphi(x,\ y,\ \lambda_{\mathrm{R,G,B}}) - \pi\}] \quad (7.7)$$

$$I_2(x,\ y,\ \lambda_{\mathrm{R,G,B}}) = I_0(x,\ y,\ \lambda_{\mathrm{R,G,B}})[1 + V\gamma(z)\cos\{\varphi(x,\ y,\ \lambda_{\mathrm{R,G,B}}) - \pi/2\}] \quad (7.8)$$

$$I_3(x,\ y,\ \lambda_{\mathrm{R,G,B}}) = I_0(x,\ y,\ \lambda_{\mathrm{R,G,B}})[1 + V\gamma(z)\cos\{\varphi(x,\ y,\ \lambda_{\mathrm{R,G,B}})\}] \quad (7.9)$$

$$I_4(x,\ y,\ \lambda_{\mathrm{R,G,B}}) = I_0(x,\ y,\ \lambda_{\mathrm{R,G,B}})[1 + V\gamma(z)\cos\{\varphi(x,\ y,\ \lambda_{\mathrm{R,G,B}}) + \pi/2\}] \quad (7.10)$$

$$I_5(x,\ y,\ \lambda_{\mathrm{R,G,B}}) = I_0(x,\ y,\ \lambda_{\mathrm{R,G,B}})[1 + V\gamma(z)\cos\{\varphi(x,\ y,\ \lambda_{\mathrm{R,G,B}}) + \pi\}] \quad (7.11)$$

The white light interferograms with five-step phase shift are recorded using a color CCD camera and stored in the computer for further processing. The white light interferograms are further digitally decomposed into its R, G and B color interferograms, thus generating three sets of phase-shifted interferograms while using only one set of experimentally recorded interferograms. Using a phase-shifting algorithm, the phase maps for RGB colors are computed and from these various parameters are extracted for biological cells. In phase-shifting interfereometry (PSI) a phase shift is introduced between the sample and the reference arm with time by moving the reference mirror sequentially in the *z*-direction using PZT and an exact phase shift of π/2 is introduced for a particular wavelength only, e.g., in the present case $\lambda_G = 532$ nm. To retrieve the phase map of the specimen, the five-step phase-shifting algorithm is used as it gives better results compared to three-step or four-step phase-shifting method and the expression for the phase map reconstruction can be given as [9–12]:

$$\phi\ (x,\ y,\ \lambda_G\ ) = \mathrm{atan}\left(\frac{2\ (I_2(x,\ y) - I_4(x,\ y))}{I_1(x,\ y) - 2I_3(x,\ y) + I_5(x,\ y)}\right) \quad (7.12)$$

where $I_1(x, y)$, $I_2(x, y)$, $I_3(x, y)$, $I_4(x, y)$, $I_5(x, y)$ are the interferograms recorded by CCD camera with five-step phase shift introduced in between, and this equation can be used for a single wavelength only, i.e., green for which the phase shift is exactly π/2, and for the other two wavelengths the phase shift δ is modified by $\delta = \frac{\lambda_i}{\lambda_G}\frac{\pi}{2}$; I = R, B and an additional term of sin δ is introduced in the numerator of the above equation [9–12]

$$\phi\ (x,\ y,\ \lambda_{R,\ B}) = \mathrm{atan}\left(\frac{2\sin\delta\ (I_2(x,\ y) - I_4(x,\ y))}{I_1(x,\ y) - 2I_3(x,\ y) + I_5(x,\ y)}\right) + \delta \quad (7.13)$$

This computational correction is equivalent to recalibrating the PZT for specific wavelengths. In fact, recalibrating the PZT from time-to-time for specific wavelengths is more challenging because the instrument always suffers from surrounding vibration. Therefore, recording white light interferogram with color camera and then post-processing with computational correction is much safer. When using $\alpha = \frac{\pi}{2}$ for the phase shift, equation (7.12) becomes independent of phase-shift calibration errors, and it becomes [9–12]:

$$\phi(x, y, t, \lambda_{R,G,B}) = \tan^{-1}\left[\frac{2\left(I_4(x, y, t, \lambda_j) - I_2(x, y, t, \lambda_j)\right)}{I_1(x, y, t, \lambda_j) - 2I_3(x, y, t, \lambda_j) + I_5(x, y, t, \lambda_j)}\right] \tag{7.14}$$

It can be seen from equation (7.15) that the phase map is related to the refractive index distribution, local height map of the specimen and the wavelength of the light source, and is given as [9–12]

$$\phi(x, y) = \frac{2\pi}{\lambda_{R,G,B}} *[2h_s(x, y)*(n_s(x, y) - n_0)] \tag{7.15}$$

where $h_s(x, y)$ is the 2D-height map of the specimen, $n_s(x, y)$ and n_0 are the refractive indices of the specimen and the surrounding medium, respectively. The phase map is determined accurately experimentally and if the geometrical height of the specimen is known then we can determine the refractive index distribution of the specimen with the prior knowledge of the refractive index of the surrounding medium.

7.2.1 Hilbert transform method for phase extraction

The phase-shifting method requires a minimum of four-step or five-step phase-shifted interferograms for reconstructing the phase map. Recently, there have been methods in which only a single interferogram is required to reconstruct the phase map, such as Hilbert phase microscopy (HPM) or Fourier phase microscopy. In these methods a single interferogram with high spatial carrier frequency is recorded using an off-axis interferometer which can be expressed as [30, 31]:

$$\begin{aligned} I(x, y, \lambda_{R,G,B}) = {} & I_R(x, y, \lambda_{R,G,B}) + I_O(x, y, \lambda_{R,G,B}) \\ & + 2[I_R(x, y, \lambda_{R,G,B})I_O(x, y, \lambda_{R,G,B})]^{1/2} \times \\ & \cos[2\pi/\lambda_{R,G,B}\, \Delta d(x, y) + \theta(x, y, \lambda_{R,G,B})] \end{aligned} \tag{7.16}$$

where $I_R(x,y, \lambda_{R,G,B})$ and $I_O(x, y, \lambda_{R,G,B})$ are the intensities of reference beam and the beam interrogated with the specimen, respectively, and $\Delta d(x, y)$ is the OPD between the light beams propagating via sample and reference arms of the interferometer and $\theta(x, y, \lambda_{R,G,B})$ is the tilt angle to obtain high density interferogram, and $\lambda_{R,G,B}$ is the peak wavelengths of red, green and blue colour light. The spectral interferogram expressed by equation (7.16) is Fourier transformed first, which gives three spectral components, DC (background), first order spectrum and its complex conjugate. The background is illuminated and one of the first order spectrum spatially filtered and then inverse Fourier transforms is computed to extract the sinusoidal term. The filtered sinusoidal term can be expressed as [30, 31]:

$$S(x, y, \lambda_{R,G,B}) = 2[I_R(x, y, \lambda_{R,G,B})\, I_O(x, y, \lambda_{R,G,B})]^{½} \times \cos[2\,\pi/\lambda_{R,G,B}\, \Delta d(x, y) + \theta(x, y, \lambda_{R,G,B})] \tag{7.17}$$

The complex analytic signal associated with the real function of $S(x,y, \lambda_{R,G,B})$ can be obtained as follows:

$$Z(x, y, \lambda_{R,G,B}) = ½[S(x, y, \lambda_{R,G,B}) + j\,\{\text{Hilbert}\,(S(x, y, \lambda_{R,G,B}))\}] \tag{7.18}$$

Using equation (7.18), the wrapped phase map associated with the complex analytic signal $Z(x, y, \lambda_{R,G,B})$ is calculated as[28]

$$\Phi(x, y, \lambda_{R,G,B}) = \tan^{-1}\{\text{Im}[Z(x, y, \lambda_{R,G,B})]/\ Re[(x, y, \lambda_{R,G,B})]\} \tag{7.19}$$

The HPM technique is highly useful for computing a phase map of dynamic samples, such as cell membrane fluctuations, cell growth, cell death and microfluidic environment. This is because it requires a single interferogram to reconstruct the phase map and hence off-axis interferometers can be used.

7.2.2 Experimental details of white light interferometry

A white light interference microscope can be realized using either a Mirau-interferometric objective lens or Michelson-type interferometric objective lens. Among these two, the Mirau-interferometric objective is most frequently used because it is basically a compact and common-path interferometer that gives simultaneous imaging of biological samples and produces white light interferograms. Inside the objective lens, a beam splitter is mounted which splits the incoming light into two beams, one goes to the sample and the other to the reference mirror which is also mounted inside it and both the beams are made to interfere onto the CCD camera plane and the phase information of the sample is encoded in that interferogram. The entire objective lens comprising lenses, reference mirror and the beam splitter is attached to a PZT which is controlled by a power controller and introduces phase shifts between the sample and the reference arm and the phase-shifted white light interferograms are recorded by a color CCD camera. A schematic diagram of a phase-shifting white light interferometric microscope (PS-WLIM) is shown in figure 7.2. The divergent white light beam from a halogen lamp (12 V50 W, LV-LH50PC) is collected and directed towards the aperture plane by the condenser lens L_1 which is further collimated by lens L_2. The collimated light beam from L_2 is passed through the field-stop (F-stop) and collected by lens L_3 which focuses the light beam on the back pupil of the Mirau objective lens for the uniform illumination at the sample plane. The aperture-stop (A-stop) and field-stop (F-stop) are used to control the spatial coherence and field of view of the system by varying their sizes, respectively. A PZT (NV40 3CLE) has been connected with the Mirau objective lens to generate equal temporal phase shift between the recorded interferograms.

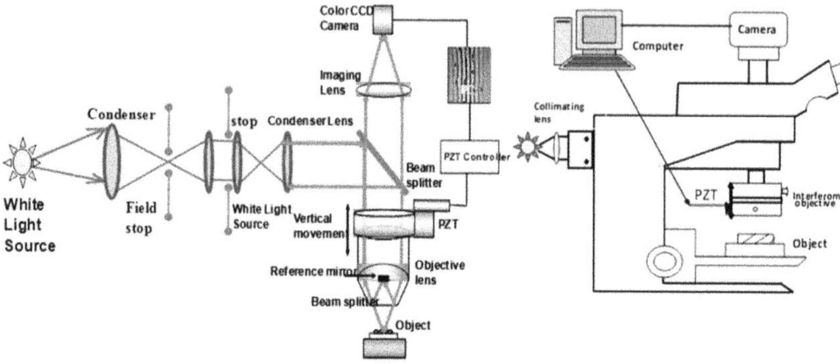

Figure 7.2. Schematic diagram illustrating phase-shifting white light interference microscopy.

The back-reflected light from the sample and reference arm get interfered inside the Mirau objective lens and recorded by a color camera (INFINITY2) by using the tube lens L_4. The recorded five-phase-shifted data frames are further utilized for the reconstruction of the phase of the objects by using a standard five-step phase-shifting algorithm.

7.3 Color fringe analysis

One of the most important advantages of WLI is that we can perform color fringe analysis if the interferogram is recorded using a color CCD camera. Secondly, the WLI (low coherence) is free from speckle noise and spurious fringes. In addition, it uses phase-shifting interferometry which is the most accurate phase-measuring technique. Figure 7.3(a) shows a single white light interferogram recorded using a color CCD camera and the interferograms corresponding to red, green and blue wavelengths can be extracted from it digitally in MATLAB.

7.4 Quantitative information about biological samples

7.4.1 Quantitative phase microscopy of heathy and anaemic blood samples

White light phase-shifting interferometry was utilized for computing the 2D phase maps of healthy and anaemic red blood cells [32]. Figure 7.4 shows the five-phase-shifted white light interferograms of normal healthy red blood cells recorded with a white light interferometric set-up with Mirau objective lens by translating the reference mirror using a PZT to introduce a particular phase shift between the sample and the reference arm using the experimental set-up in figure 7.2.

Each white light interferogram was decomposed into three primary colours (R, G, and B), thus three sets of five- step phase-shifted interferograms for each colour are generated while conducting the experiment only once. A five- step phase-shifting algorithm was used to reconstruct the phase map for each colour using equation (7.14). The figure 7.5(a) white light interferogram and figures 7.5(b), (c) and (d) show

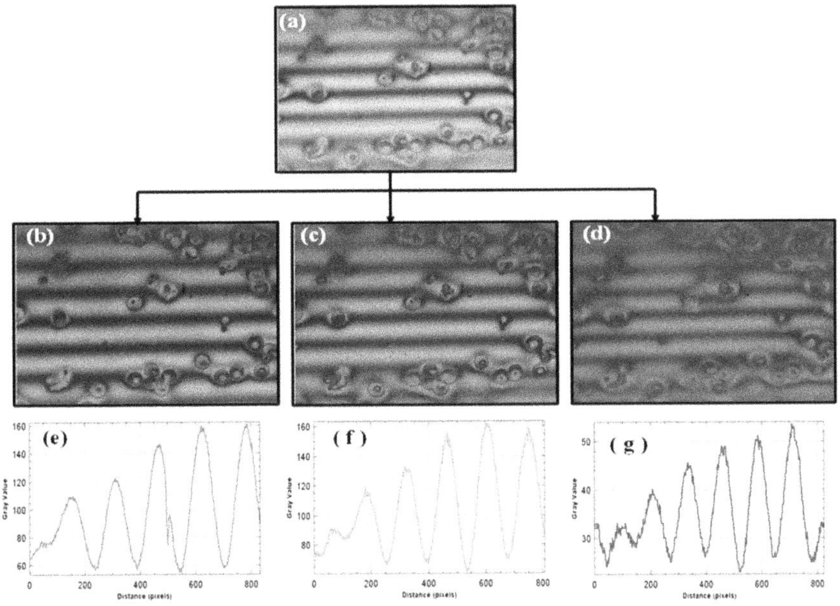

Figure 7.3. (a) White light interferogram, (b), (c), (d) interferograms with red, green, and blue color light and (e), (f) and (g) the corresponding line profiles for three colors. Panels (b), (c) and (d) show the interferograms corresponding to red, green and blue colour light digitally decomposed using MATLAB code. Different wavelengths lead to different interferogram periods which can be seen clearly from their line profiles shown below (e), (f) and (g), respectively.

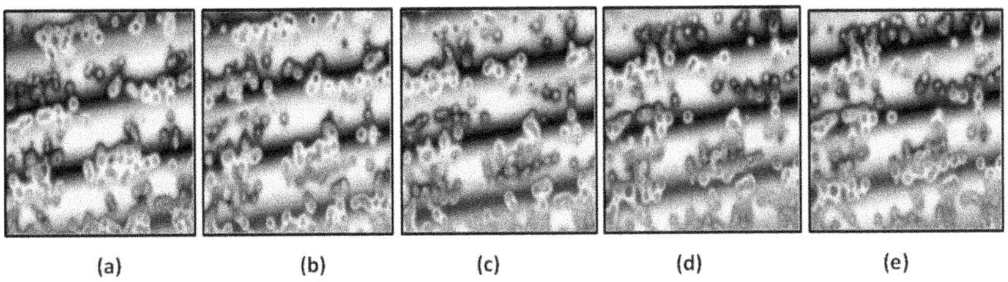

Figure 7.4. White light interferograms of red blood cells using five-step phase-shifting method; (a) $-\pi$, (b) $-\pi/2$, (c) 0, (d) $\pi/2$ and (e) π phase-shifted interferograms.

the reconstructed phase profiles of normal red blood cells corresponding to red, green and blue wavelengths, which can be retrieved by using any of the phase-shifting algorithms.

Anaemia can be considered as a health condition of deficiency of healthy red blood cells in our body, and can lead to the abnormal shape and size of these cells, which can be easily tracked by analysing their phase profiles. Figure 7.6 shows the phase profiles of anaemic cells which are quite different from those of normal cells. A noteworthy change can be observed in the phase profiles of anaemic cells as compared to normal ones [32].

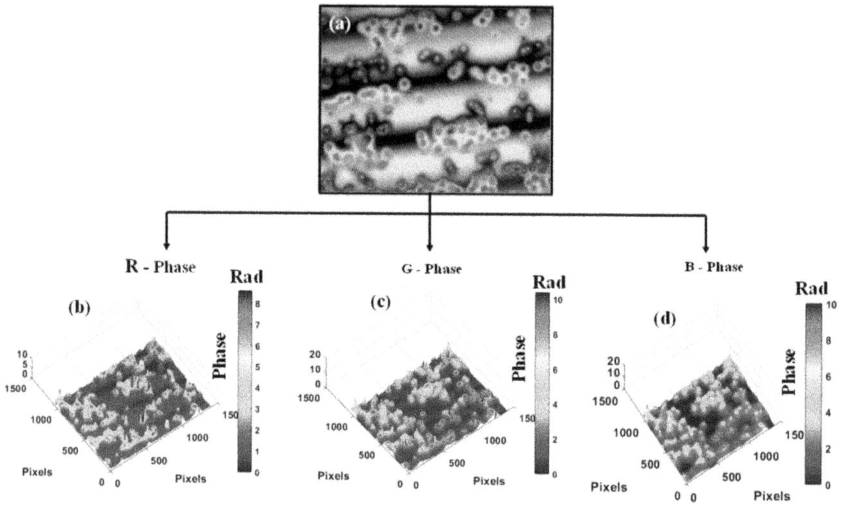

Figure 7.5. (a) White light interferogram and (b), (c) and (d) phase profiles of normal red blood cells corresponding to red, green and blue wavelengths, respectively (colour bar represents phase in rad).

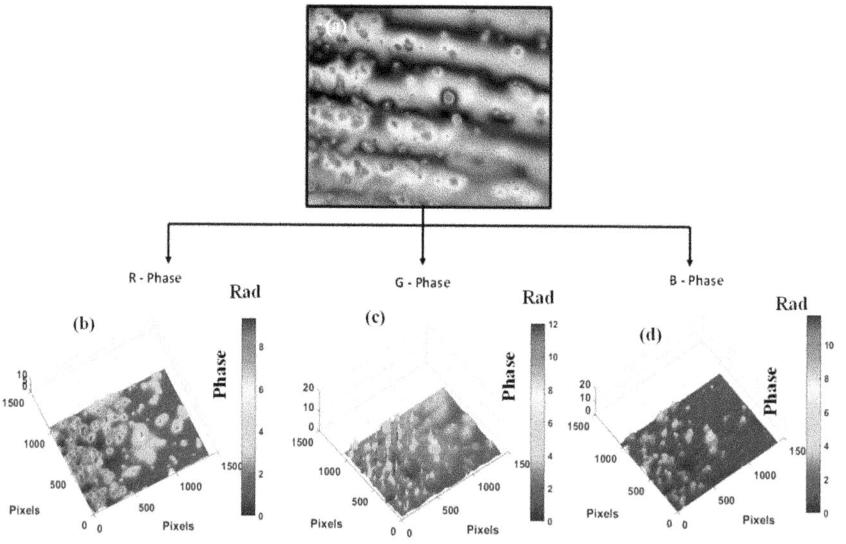

Figure 7.6. (a) Single white light interferogram, (b), (c) and (d) phase profiles of anemic red blood cells corresponding to red, green and blue wavelengths (colour bar represents phase in rad).

QPM techniques have become an important tool for the determination of various biophysical parameters of cells and tissues. The 3D phase map of the cell is reconstructed using QPM and equation (7.15). Once the phase map is reconstructed carefully, one can determine the following parameters using the phase value.

(a) Cell morphology: The cellular morphology and topography can be studied and measured using phase images in QPM techniques. It can be seen from figures 7.5 and 7.6 above that the morphology (shape) of the RBCs is different for normal and anaemic cases. Further, the phase delay in QPM can be used to calculate the height parameter of RBCs. This technique has important clinical implications like in malaria where parasite *plasmodium falciparum* causes biochemical, structural and mechanical changes to host RBCs [9, 12, 17], which can be determined using QPM. The major structural changes result in loss of cell volume and change in the appearance of membrane surface, which can also be observed in QPM [10, 11].

(b) Determination of refractive index of RBCs: Another important parameter in studying RBCs is the determination of refractive index as a function of wavelength of light using WLI, details of which can be found elsewhere [11, 18–25]. The hemoglobin concentration in the erythrocytes is mainly responsible for its refractive index and any change in this value can alter optical properties of blood [8]. The major source of scattering of light in whole blood is the refractive index mismatch between erythrocytes and blood plasma [18–25]. The Barer's expression for refractive index of RBC is

$$\Delta n(x, y) = \alpha\,(\lambda)\; C(x, y) \;+\; nX(x, y, \lambda) \tag{7.20}$$

where $\Delta n(x,y)$ is refractive index distribution of RBC, $\alpha(\lambda)$ is the incremental refractive index, $C(x, y)$ is cytoplasmic Hb concentration in the RBC and $n_X(x, y, \lambda)$ is the average refractive index of other molecules besides Hb in RBC [18–25]. One of the important applications of QPM could be the monitoring of hemolysis when hemoglobin leaks out from erythrocytes and, consequently, the local concentration of hemoglobin rapidly changes. Barer's expression assumes that the real part of the refractive index, n, is proportional to the hemoglobin concentration, C which is given as $n = n_{H_2}O + \alpha C$, where $n_{H_2}O$ is the refractive index of distilled water in which the cells are floating, α is the specific refractiive index increment and C is the concentration [18–25]. Since the phase is determined using equation (7.15) with white light phase-shifting interferometry and a five-step phase-shifting algorithm and if the geometrical thickness or height map $h(x, y)$ of the cell and the refractive index of the sounding medium is known, then the refractive index distribution of RBC can be determined using the phase map of the cell using the following equation [18–26]:

$$n_{\text{cell}}(x, y, \lambda) = \frac{\Delta\varphi(x, y, \lambda)}{2\pi h(x, y)} + n_{\text{medium}} \tag{7.21}$$

In general, the refractive index and geometrical thickness of the biological sample are coupled, therefore, the alternative method of determining the refractive index is the de-coupling between geometrical thickness and refractive index of phase object [18]. One of the possible methods to perform this is by means of preparing the sample in two media of known refractive indices and performing the experiment twice and then determining two phase maps separately, for medium of known refractive index, n_1, and for medium of known refractive index, n_2 [18, 25] and then solving the coupled equation as follows;

$$n_s(x, y) = \frac{\lambda\Delta\phi_1(x, y)}{2\pi \text{h}(x, y)} + n_1 = \frac{\lambda\Delta\phi_2(x, y)}{2\pi \text{h}(x, y)} + n_2 \tag{7.22}$$

Rearranging equation (7.22), the thickness '$h(x, y)$' of the phase object or the specimen can be computed by the following expression [18]:

$$h(x, y) = \frac{\lambda(\Delta\phi_1(x, y) - \Delta\phi_2(x, y))}{2\pi(n_2 - n_1)} \tag{7.23}$$

(c) Dry mass: It consists mainly of cell protein and can be quantified using QPI technique [13]. Dry mass is a reliable indicator of cell growth and divison as it is independent of intra-cellular water content [13]. A known relationship between refractive index and dry mass contents is used to convert phase image into dry mass. Calculation of dry mass in the cell is done by using relation [18–26]:

$$\sigma(x, y) = \frac{\lambda}{2\pi\alpha(x, y)}\phi(x, y) \tag{7.24}$$

where $\alpha(x, y)$, $\Phi(x, y)$ are refractive index increment and phase map, respectively. In the life cycle of cell, growth and I are important parameters which are genteically controlled according to cell type. Loss of this regulation results in uncontrolled proliferation of nonfuntional abnormal cells. These parameters can also be measured using phase-shifting white light interference microscopy [18–26].

(d) Cell dynamics: The RBC membrane is a bilayered structure consisting of proteins and lipids, which helps in deformation and stability of cell structure during physiological functions. The dynamic nature of cell membrane is seen in a phenomenon called 'flickering' and any genetic alteration results in sickle cell disease [27–31].

7.4.2 Quantitative phase imaging of normal and cancerous cells

White light phase-shifting interferometry can also be utilized for reconstructing the phase distribution of normal and cancerous breast cells. Samples for normal and cancerous were obtained from AIIMS Hospital Delhi. This *ex-vivo* study was performed at Bio-photonics Laboratory, Indian Institute of Technology (IIT) Delhi. The ethical clearance was taken from the ethics committee of IIT Delhi and All India Institute of Medical Science (AIIMS) for performing experiments on cancer tissues with the permission of patients. The breast tissue samples were collected from 16 different patients during surgical breast biopsy at AIIMS, Delhi, India. Samples were fixed in formalin and stored at −40 °C [16]. Thick normal and tumour tissue samples were cut into thin slices (<5 μm) using microtome and prepared on the reflecting microscopic quartz glass slides for conducting the experiments. Experiments for imaging were conducted soon after microtome sectioning, with the surrounding temperature 22 °C. Similarly, classification between normal and cancerous cells can be employed using PS-WLIM. Samples were prepared on clean glass substrates and white light interference images recorded using color CCD. Five-step phase-shifted white interferograms were recorded using the experimental set-up in figure 7.2. The red (R), green (G) and blue (B) frames are digitally separated using MATLAB software and the 2D and 3D phase maps are reconstructed using a five-phase- shifted algorithm. The 2D phase maps of R, G, B frames for normal breast cells are shown in figures 7.7(b), (c), (d) and corresponding 3D phase maps are shown in figures 7.7(b′), (c′), and (d′), respectively. Visible light is used to image the

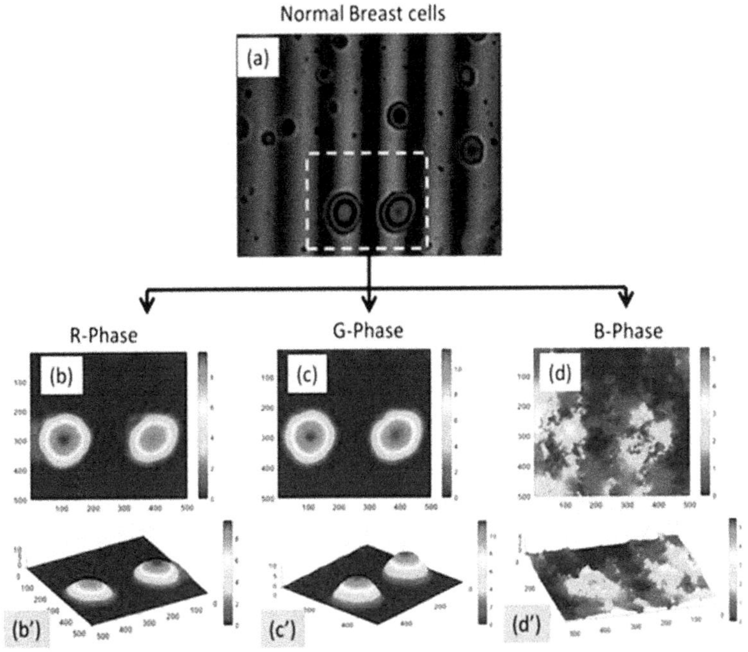

Figure 7.7. (a) White light interferogram of normal breast cells, (b) 2D phase map of red frame, (b′) 3D phase map of red frame, (c) 2D phase map of green frame, (c′) 3D phase map of green frame, (d) 2D phase map of blue frame and (d′) 3D phase map of blue frame.

biological specimen, but high-energy light ranging from 400 to 450 nm in the visible spectrum is absorbed more compared to R, and G.

Therefore, the quality phase maps for R and G channels are good, whereas B channel represents a poor quality phase map. The blue signal is already weak due to absorption of light by the biological specimen, which further degrades using the single-chip CCD camera, however, improvement in signal intensity could be obtained using a three-chip colour camera. The experiment was conducted for cancerous cells (ductal carcinoma) in the same way as above. The 2D phase maps of cancerous breast cells are shown in figures 7.8(a), (b), (c) and the corresponding 3D phase maps are shown in figures 7.8(b′), (c′), and (d′), respectively. The normal and cancerous breast cells can be distinguished as there is an increment in the phase value for cancerous cells as compared to the normal cells. Thus, QPM could be an effective tool for the diagnoses of cancerous cells.

Once the phase map is measured then one can determine the shape, morphology, the refractive index and dry mass of cancerous cells and tissues using white light interferometry. Further, this technique is label-free and also free from speckle as low coherence (white) light used in this technique. Further, the interferometer system is common- path and both imaging and interferometry can be done simultaneously, therefore, it has great potential for the classification of cancerous cells [16, 31–35].

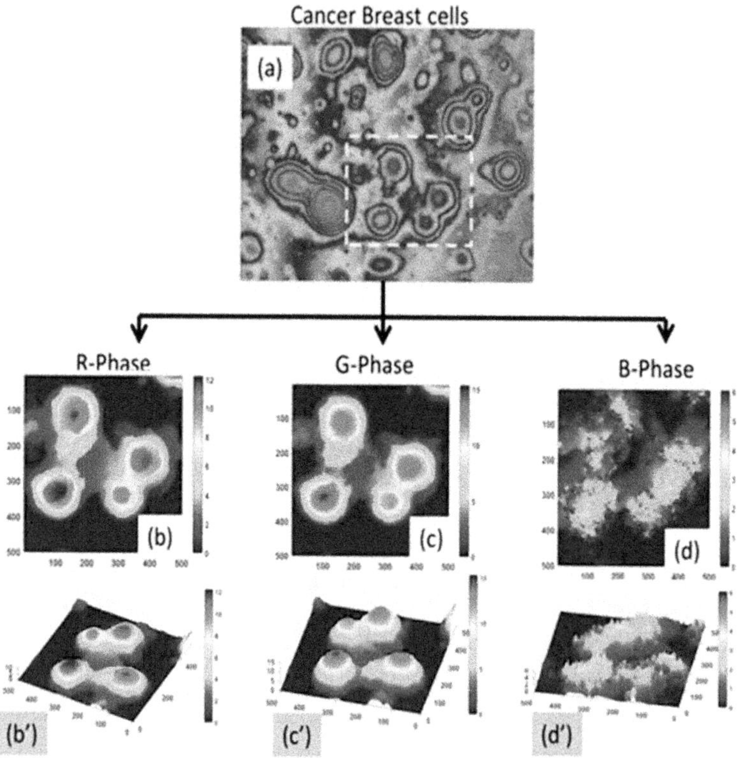

Figure 7.8. (a) White light interferogram of cancerous breast cells, (b) 2D phase map of red frame, (b′) 3D phase map of red frame, (c) 2D phase map of green frame, (c′) 3D phase map of green frame, (d) 2D phase map of blue frame and (d′) 3D phase map of blue frame.

7.4.3 Quantitative phase imaging of *Escherichia coli* bacteria

Finally, low-coherence (white light) interference microscopy with colour fringe analysis was also tested using *E. coli* bacteria as a sample. The white light interferogram of the *E. coli* bacteria is shown in figure 7.9(a). The R, G and B channels are digitally filtered, and phase maps are obtained. The 2D phase maps of the *E. coli* bacteria for R, G and B channels are shown in figures 7.9(b), (c) and (d), respectively. The corresponding 3D phase maps are shown in figures 7.9(b′), (c′) and (d′), respectively. In summary, one can used phase-shifting WLI with colour fringe analysis for label-quantitative phase imaging of bacteria samples. One can determine, the shape, morphology and the refractive index of these biological samples. Therefore, this technique has great potential for label-free detection and diagnosis of the various diseases [31–36].

In summary, white light phase-shifting interferometry has great potential for label-free QPM of biomedical applications. This technique is speckle-free and no spurious fringes occur. Hence it is a highly accurate phase-measurement technique compared to laser based techniques. Further, using this technique the refractive

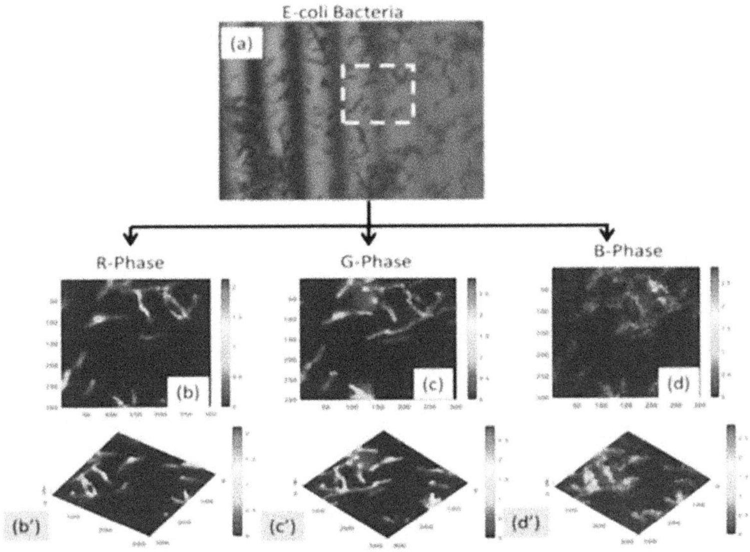

Figure 7.9. (a) White light interferogram of *E. coli* bacteria,(b) 2D phase map of red frame, (b′) 3D phase map of red frame, (c) 2D phase map of green frame, (c′) 3D phase map of green frame, (d) 2D phase map of blue frame and (d′) 3D phase map of blue frame.

index of the samples as a function of wavelength can be determined. In addition, we can also determine various other biophysical parameters.

References

[1] Mandel L and Wolf E 1965 Coherence properties of optical fields *Rev. Mod. Phys.* **37** 231

[2] Bouma B E and Tearney G J 2002 *Handbook of Optical Coherence Tomography* (New York: Marcel Dekker, Inc)

[3] Drexler W and Fujimoto J G 2008 *Optical Coherence Tomography Technology and Applications* 1st edn (New York: Springer)

[4] Fercher A F, Mengedoht K and Werner W 1988 Eye-length measurement by interferometry with partially coherent-light *Opt. Lett.* **13** 186–8

[5] Fercher A F, Hitzenberger C and Juchem M 1991 Measurement of intraocular optical distances using partially coherent laser-light *J. Mod. Opt.* **38** 1327–33

[6] Hitzenberger C K 2017 *Low-Coherence Interferometry, Handbook of Visual Optics* (Boca Raton, FL: CRC Press)

[7] Häusler G and Lindner M W 1998 Coherence radar and spectral radar—new tools for dermatological diagnosis *J. Biomed. Opt.* **3** 21–31

[8] Hitzenberger C K, Baumgartner A, Drexler W and Fercher A F 1994 Interferometric measurement of corneal thickness with micrometer precision *Am. J. Ophthalmol.* **118** 468–76

[9] Hitzenberger C K, Baumgartner A, Drexler W and Fercher A F 1999 Dispersion effects in partial coherence interferometry: implications for intraocular ranging *J. Biomed. Opt.* **4** 144–51

[10] Mehta D S and Srivastava V 2012 Quantitative phase imaging of human red blood cells using phase-shifting white light interference microscopy with colour fringe analysis *Appl. Phys. Lett.* **101** 203701

[11] Pham H, Bhaduri B, Huafeng D and Popescu G 2012 Spectroscopic diffraction phase microscopy *Opt. Lett.* **37** 3438–40

[12] Zhernovaya O, Sydoruk O, Tuchin V and Douplik A 2011 The refractive index of human hemoglobin in the visible range *Phys. Med. Biol.* **56** 4013–7

[13] Mehta D S 2018 Quantitative phase optical microscopic techniques for biomedical imaging and diagnostic applications *Proc. Natl. Acad. Sci., India, Sect. A Phys. Sci.* **88** 437–51

[14] Native A and Shaked N T 2017 Compact interferometric module for full-field interferometric phase microscopy with low spatial coherence illumination *Opt. Lett.* **42** 1492–4

[15] Bhaduri B, Edwards C, Pham H, Zhou R, Nguyen T H, Goddard L L and Popescu G 2014 Diffraction phase microscopy: principles and applications in materials and life sciences *Opt. Lett.* **37** 1094–6

[16] Dubey V, Singh G, Singh V, Ahmad A and Mehta D S 2016 Multi-spectral quantitative phase imaging of human red blood cells using inexpensive narrow band multi-color LEDs *Appl. Opt.* **55** 2521–5

[17] Dubey V, Ahmad A, Butola A, Qaiser D, Srivastava A and Mehta D S 2019 Low coherence quantitative phase microscopy with machine learning model and Raman spectroscopy for the study of breast cancer cells and their classification, *Appl. Opt.* **58** A112–9

[18] Singh V, Srivastava V and Mehta D S 2020 Machine learning-based screening of red blood cells using quantitative phase imaging with micro-spectro-colorimetry *Opt. Laser Technol.* **124** 105980

[19] Cardenas N and Mohanty S 2013 Decoupling of geometric thickness and refractive index in quantitative phase microscopy *Opt. Lett.* **38** 1007–9

[20] Park Y K, Popescu G, Lykotrafitis G, Choi W, Feldand M S and Suresh S 2008 Refractive index maps and membrane dynamics of human red blood cells parasitized by *Plasmodium falciparum Proc. Natl Acad. Sci.* **105** 13730–5

[21] Popescu G, Ikeda T, Dasari R R and Feld M 2006 Diffraction phase microscopy for quantifying cell structure and dynamics *Opt. Lett.* **31** 775–7

[22] Zhernovaya O, Sydoruk O, Tuchin V and Douplik A 2011 The refractive index of human hemoglobin in the visible range *Phys. Med. Biol.* **56** 4013–17

[23] Jang Y, Jang J and Park Y 2012 Dynamic spectroscopic phase microscopy for quantifying hemoglobin concentration and dynamic membrane fluctuation in red blood cells *Opt. Express* **20** 9673–81

[24] Park Y K, Diez-Silva M, Popescu G, Lykotrafitis G, Choi W, Feld M S and Suresh S Refractive index maps and membrane dynamics of human red blood cells parasitized by Plasmodium falciparum *Proc. Natl Acad. Sci.* **105** 13730–5

[25] Popescu G, Park Y K, Lue N, Catherine B, Deflores L, Dasari R R, Feldand M S and Badizadegan K 2008 Optical imaging of cell mass and growth dynamics *Am. J. Physiol.-Cell Physiol.* **295** 538–43

[26] Timothy R, Higgins J M, Diez-Silva M, Peng Z, Dao M, Dasari R R, Suresh S and Park Y K 2012 Optical measurement of biomechanical properties of individual erythrocytes from a sickle cell patient *Acta Biomater.* **8** 4130–8

[27] Shaked N T, Satterwhite L L, Telen M J, Truskey G A and Wax A 2011 Quantitative microscopy and nanoscopy of sickle red blood cells performed by wide field digital interferometry *J. Biomed. Opt.* **16** 030506

[28] Rappaz B, Cano E, Colomb T, Kohn J, Simanis V, Magistretti P J, Marquet P and Depeursinge C 2009 Noninvasive characterization of the fission yeast cell cycle by monitoring dry mass with digital holographic microscopy *J. Biomed. Opt.* **14** 34049–56

[29] Kim S, Kim Y, Cho S and Park Y K 2012 Optical imaging techniques for the study of malaria *Trends Biotechnol.* **30** 71–9

[30] Popescu G, Park Y, Choi W, Dasari R, Feld M and Badizadegan K 2008 Imaging red blood cell dynamics by quantitative phase microscopy *Blood Cells Mol. Dis* **41** 10–6

[31] Ikeda T, Popescu G, Dasari R R and Feld M S 2005 Hilbert phase microscopy for investigating fast dynamics in transparent systems, *Opt. Lett.* **30** 1165–7

[32] Srivastava V, Anna T and Mehta D S 2012 Single shot quantitative imaging of biological cells using low coherence off-axis interferometric Hilbert phase microscopy *J. Opt.* **14** 125707

[33] Mann P, Singh V, Tayal S, Thapa P and Mehta D S 2022 White light phase shifting interferometric microscopy with whole slide imaging for quantitative analysis of biological samples *J. Biophoton.* **15** e202100386

[34] Kong K, Zaabar F, Rakha E, Ellis I, Koloydenko A and Notinghe I 2014 Towards intra-operative diagnosis of tumours during breast conserving surgery by selective-sampling Raman micro-spectroscopy *Phys. Med. Biol.* **59** 6141–54

[35] Huang S, Yang J, Fong S and Zhao Q 2020 Artificial intelligence in cancer diagnosis and prognosis: opportunities and challenges *Cancer Lett.* **471** 61–71

[36] Butola A, Ahmad A, Dubey V, Srivastava V, Qaiser D, Srivastava A, Senthilkumaran P and Mehta D S 2019 Volumetric analysis of breast cancer tissues using machine learning and swept-source optical coherence tomography *Appl. Opt.* **58** A135–41

Chapter 8

Artificial intelligence: a computational tool to interpret quantitative phase imaging

Chapter 8 presents the unique integration of quantitative phase microscopy and artificial intelligence to deal with the various challenging problems in biomedical applications. This chapter covers the need of machine learning and deep learning for the interpretation of phase microscopy followed by the fundamental principle of convolutional neural network. In addition, various applications of conventional and modern machine learning architecture are shown to improve the classification accuracy, resolution, space bandwidth and chemical specificity of phase microscopy.

8.1 Introduction

Quantitative phase imaging (QPI) deals with the 'phase information' of specimens, which is considered a fundamental problem in optical imaging. Since it was first invented two decades ago, QPI has been developed both computationally and experimentally to improve the imaging output and is widely used for various biological applications. In particular, the last decade has witnessed an unprecedented number of applications by combining two fundamental research fields, i.e., QPI and artificial intelligence (AI) [1–5]. While the two research fields are well established now, the elementary objectives were proposed decades ago. Further, the practical implementations have been made possible only after significant developments in image sensors and the graphics processing unit (GPU). With these advancements, integration of QPI with AI has attracted many researchers to explore the more exciting and challenging problems in biomedical applications.

The fast and label-free nature of QPI has become extremely popular in optical imaging, and AI assigns an artificial system to mimic human intelligence such as classification between two closely alike objects and differentiation/segmentation of the object based on abstract features. In particular, the subfield of AI called machine learning (ML) is a data driven approach explicitly designed and optimized by fine

doi:10.1088/978-0-7503-3987-2ch8

turning the hyper-parameters to reflect certain features through learning rules. The modern ML approach called deep neural network (DNN) or convolutional neural network (CNN), inherently supports abstract characterization of the physical features, i.e., scattering and absorption coefficient, refractive index, geometric features, as well as relevant statistical features. Using CNN is especially useful when processing large-scale and complex datasets that obstruct human investigation [1–5].

8.2 Machine learning to understand QPI/need of machine learning and deep learning in phase microscopy

QPI measures the incident field's phase shift induced by the sample using both interferometry and non-interferometry based techniques. QPI techniques have been developed to recover the sample phase map in both 2D and 3D. Fast acquisition and utilizing intrinsic contrast of the object enables quantitative measurements of different biophysical parameters at cellular and sub-cellular level. For example, surface area, volume, dry mass, membrane fluctuations can be measured for individual cells. Although QPI is a unique and powerful method to measure the cell morphology and dynamics, the chemical specificity of intrinsic refractive index contrast-based imaging is limited compared to the conventional direct labelling technique such as fluorescence microscopy. The primary bottleneck comes with mapping chemical composition with the phase image of the object. ML is a powerful tool to learn abstract features from the phase image for the classification purpose and can learn the mapping between fluorescence and QPI to learn chemical specificity using a data driven approach.

An advanced ML technique, i.e., CNN is being adopted for various applications such as classification and segmentation of biomedical images. A popular approach is to perform transfer learning, i.e., use a pretrained CNN (ResNet, AlexNet, VGG-16, Inception etc) and retrain the weights of the CNN for a biomedical dataset. It has been shown that optical coherence tomography (OCT)-assisted CNN can be useful in different applications such as detection of ocular disease, predicting diabetic macular edema, choroidal segmentation and early onset glaucoma [1–5]. Also, other biological applications of OCT+CNN have been highlighted recently, such as denoise of the optic nerve head, classification of human atrial tissue, plaque characterization in intracoronary OCT images, among others [6–8]. Additionally, it has also opened new possibilities in QPI such as rapid optical screening of anthrax spores, phase unwrapping, digital phase staining, bright field holography, virtual staining, improved resolution and others [9–12].

8.2.1 Convolutional neural network architecture

CNN consists of a set of convolutional layers followed by other layers, i.e., maxpooling, rectifier linear unit (ReLU)/leaky ReLU and a fully connected layer, followed by a soft-max layer for classification purpose. Figure 8.1 depicts a schematic of layers presented in CNN. Learning of the network starts with the

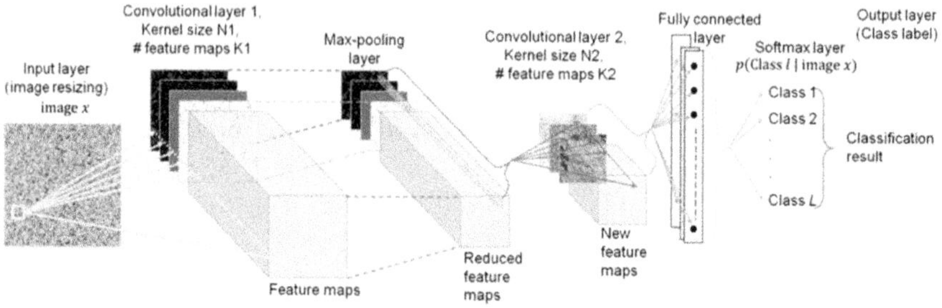

Figure 8.1. Schematic diagram of a CNN consisting of convolutional layer of different kernel size and feature map, maxpooling, fully connected layer, SoftMax and output layer. Reproduced with permission from OSA [3], copyright 2020 Optica Publishing Group.

convolution operation between input image and the first convolutional layer. In general, convolution is an operation between two functions (*f* and *k*) that produces a third function '*s*' expressing the modification in shape of one function by the other. Mathematically [9–12],

$$s(t) = \int f(x)k(t - x)dx \tag{8.1}$$

Equation (8.1) represents 1D convolution operation and is denoted as $f * k$. In CNN terminology, the first term '*f*' refers the input image and the second term k is a kernel. The discrete convolution of equation (8.1) can be defined as:

$$s(i) = f * k(i) = \sum_{m} f(m)k(i - m) \tag{8.2}$$

Equation (8.2) shows the 1D discrete convolution. However, in ML, the thousands of input images correspond to multidimensional array and similarly kernels are a multidimensional array of parameters adapted by the learning algorithm. For example, if '*f*' and '*k*' represent the 2D input image and kernel, respectively, the convolution operation can be written as

$$s(i, j) = (f * k)(i, j) = \sum_{m}\sum_{n} f(m, n)k(i - m, j - n) \tag{8.3}$$

Using the commutative property of convolution, equation (8.3) becomes

$$s(i, j) = (k * f)(i, j) = \sum_{m}\sum_{n} f(i - m, j - n)k(m, n) \tag{8.4}$$

Equation (8.4) represents what is more straightforward to implement in an ML algorithm. However, instead of using convolution, cross-correlation is mostly preferred in neural networks. Cross-correlation is similar to convolution without flipping the kernel, i.e. [9–12]

$$s(i, j) = (k * f)(i, j) = \sum_{m}\sum_{n} f(i + m, j + n)k(m, n) \tag{8.5}$$

Thus, there is one feature map for each kernel, which is approximately the same size as the input image. The ReLU is used after the first convolution layer which clips the output so that the network remains stable. The mathematical function of the ReLU layer can be defined as

$$f(x) = \begin{cases} 0 \text{ for } x < 0 \\ x \text{ for } x \geqslant 0 \end{cases} \tag{8.6}$$

ReLU function implies that the value at each pixel is 0 if it is negative and it remains the same if equal to or more than 0. The feature map is reduced to spatially half the size in the next layer, which is the maxpooling layer. Mathematically,

$$f_{i,j} = \max\{x_{i+k-1,\, j+l-1} \forall\ 1 \leqslant k \leqslant m \text{ and } 1 \leqslant l \leqslant m\} \tag{8.7}$$

It retains the maximum feature value in a region of 2×2 pixels ($k = m = 2$), i.e., maximum activation in each local region (2×2 pixels) is retained for next layer analysis.

The next layer is again a convolutional layer which works the same as the first layer with the same or different number of kernels. Second layer identify local spatial features maps generated by the previous convolutional layers. The process repeats many times with maxpooling and ReLU layer and finally the fully connected layer combines the activations of all the neurons. A fully connected layer computes cross-feature and cross-spatial relationships to generate activations for each class of the datasets. In the fully connected layer, information from distantly located features is combined and assessed for each class. The fully connected layer is followed by a softmax layer, which contains the softmax function to compute the relative activations of all the classes. The softmax function can be defined as:

$$\text{softmax}(x(i)) = \frac{e^{x(i)}}{\sum_j e^{x(j)}} \tag{8.8}$$

where $\sum_i \text{softmax}(x(i)) = 1$.

Lastly, the output layer is used to identify the class label for which the softmax function generates the maximum activation. In short, functional architecture of a neural network can be described as a combination of multiple convolutional layers which compute texture features using a fixed number of kernels of specific size, computing the local cross-feature patterns. Finally, a fully connected layer employing cross-spatial-feature patterns is used to differentiate the datasets into their relevant classes.

8.2.2 Conventional versus deep learning approach for classification of QPI dataset

8.2.2.1 Conventional approach

The classification problem is typically divided into training and testing process. During the training of the network, the images/extracted features with specified class are used to train and optimize the classifier. This type of training is known as supervised learning where the optimization process, i.e., selection of physical/texture features and designing the loss function, plays an important role to achieve better classification accuracy. The training of the network is a tedious and time-consuming process depending on the number of weights presents in the classifier, given epochs during training and most importantly size of the training datasets. After sufficient training and minimizing the loss function, the performance of the network is evaluated on the testing datasets and corresponding ground truth. If the training datasets share common morphology, sub-cellular structure, chemical composition, and phase map, the classification approach performs even better than human investigation.

The conventional classification approach works by extracting texture, physical and morphological features from the multiple QPI images. These features extracted from the QPI images are used to train and optimize the network. It is important to note that extraction of relevant features is crucial to train and optimize the performance of classifier. The classification approach for QPI datasets used in many works since the first paper reported in 2005 by Javidi *et al* [13, 14]. After this work, QPI has been used in various classification problems such as normal and infected red blood cells (RBCs), sperm cells, stem cells, cynobacteria, macrophages, melanoma cells, particulate matter, and cancer cells [15–18]. Here, we will cover the useful morphological, physical and texture features used in the aforementioned study to differentiate between multiple classes of the datasets.

Optical thickness (OT) of the phase map can be used as an important morphological feature and can be defined in reflection mode geometry as:

$$\mathrm{OT} = \varphi(x, y) \times \frac{\lambda}{4\pi} \tag{8.9}$$

where $\varphi(x, y)$ represents the phase of the object and λ is the incident wavelength used to calculate the phase of the object. OT further can be used to calculate the volume of the object. Mathematically:

$$V = \iint \mathrm{OT}(x, y)dxdy \tag{8.10}$$

dx, dy are the calibrated pixel widths along the x and y directions.

Surface area of the phase image is used as a morphological feature and calculated by summing the area elements defined as

$$dS = \iint \sqrt{1 + g_x + g_y}\, dxdy \tag{8.11}$$

In addition, sphericity of the phase map is defined by the ratio between volume and surface area shown in equations (8.10) and (8.11).

$$\psi = \frac{4.84 \times V^{\frac{2}{3}}}{S} \tag{8.12}$$

The other important features such as surface-to-volume ratio and eccentricity can be calculated using the aforementioned formulae. On the other hand, texture feature, i.e., mean, variance, energy, entropy, kurtosis and skewness are also used in the previous study to separate the phase maps into their relevant classes. The mathematical formulation of these parameters is defined in table 8.1.

Table 8.1. Mathematical formula of the texture feature.

Feature	Formula
Mean (μ)	$\frac{1}{N}\frac{1}{M}\sum_{i=1}^{N}\sum_{j=1}^{M}\phi_{ij}$
Variance (σ^2)	$\frac{1}{N}\sum_{i=1}^{N}\sum_{j=1}^{M}\left(\phi_{ij}-\mu\right)^2$
Energy	$\sum_{i,j=0}^{N-1}\left(\phi_{ij}\right)^2$
Entropy	$-\sum_{i=1}^{N}p(\phi_i)\log_2 p(\phi_i)$
Skewness	$\frac{1}{N}\sum_{i=1}^{N}\sum_{j=1}^{M}\left[\frac{(\phi_{ij}-\mu)}{\sigma}\right]^3$
Kurtosis	$\frac{1}{N}\sum_{i=1}^{N}\sum_{j=1}^{M}\left[\frac{(\phi_{ij}-\mu)}{\sigma}\right]^4-3$

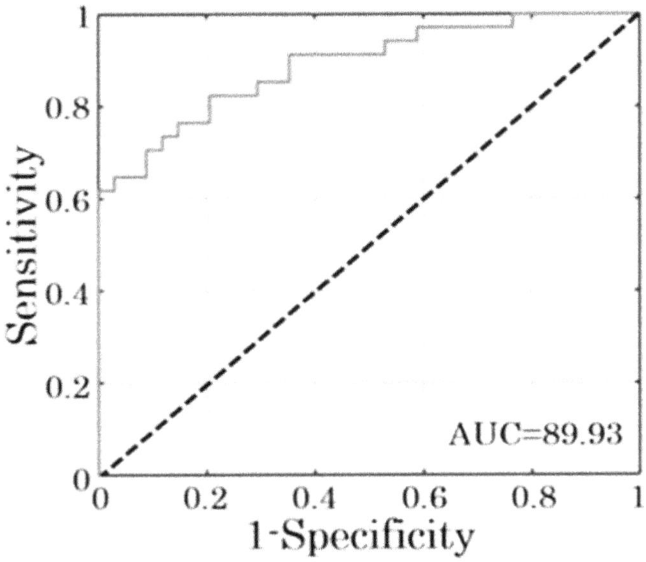

Figure 8.2. ROC curve for testing dataset for control and H_2O_2 concentration treated sperm cell. Reproduced with permission from [17], copyright 2019, The Author(s). With permission of Springer.

These morphological/texture features are used by many researchers and further used as a standard ML classifier to differentiate between two classes of the phase images. For example, Dubey *et al* [17] used these features to differentiate between normal and H_2O_2 affected sperm cells. Figure 8.2 shows the receiver operating characteristics (ROC) curve for testing a dataset of sperm heads for controlled condition and 10 μM ml^{-1} H_2O_2 concentration treated sperm cells. A total of eleven texture and morphological features are extracted to classify the datasets. The phase datasets of the cells are extracted using a partially spatially coherent quantitative phase microscope. A support vector machine (SVM)-based classifier is used to perform the classification study where 60% of the datasets are used for the training and 40% are for testing purpose. Sensitivity, specificity and accuracy of 88.61%, 91.18% and 89.93%, respectively, are found for the classification of control and oxidatively stressed sperm cells.

However, feature extraction-based classification approach does not work very well as the number of classes and the size of datasets is increased. Butola *et al* [16] used the same features to separate 9600 phase images (2400 normal, 2400 ethanol, 2400 H_2O_2 and 2400 cryopreserved) of sperm cells into their relevant classes. Three classifiers, i.e., SVM, KNN and Naive Bayes are used for the classification purpose. Figure 8.3 depicts the confusion matrix and ROC curve for SVM, Naive Bayes and KNN models. The diagonal elements of the matrix show the correct prediction by the network and off-diagonal correspond to the wrong prediction. Additionally, the ROC curve represents the accuracy of the model and is plotted between normal and other classes of sperm cells. The *x*- and *y*-axes of the ROC curve show false positive and true positive rate, respectively. It is found that the SVM, Naive Bayes and KNN models offer classification accuracy of 39.1%, 36.4% and 37.5%, respectively.

The poor classification accuracy of these models can be explained by increasing the number of classes in the datasets. Increasing the number of classes and phase images makes the datasets very complex to handle with simple ML models. The extraction of relevant morphological and texture features will certainly affect the final accuracy of the model. Diversity of diagnostic features, variation in imaging system such as illumination and numerical aperture (NA) dependent resolution, and difficulty in deriving a relevant and consistent feature create difficulty in applying feature

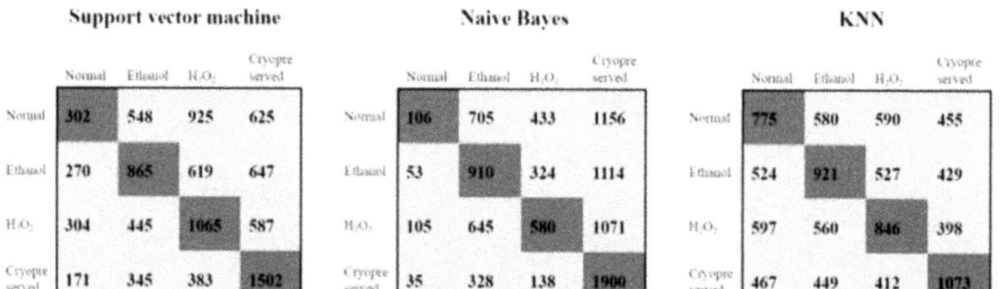

Figure 8.3. Evaluation of the performance of machine learning models for classification of 9600 phase images (2400 normal, 2400 Ethanol, 2400 H_2O_2 and 2400 cryopreserved) of head part of sperm cells. Confusion matrix and ROC curve show the classification performance and final accuracy of SVM, Naive Bayes and KNN model. SVM, Naive Bayes and KNN model provide classification accuracy of 39.1%, 36.4% and 37.5%, respectively. Reproduced with permission from [16], copyright 2020, The Author(s). With permission of Springer.

extraction-based ML techniques for classification of the cells and tissue. For example, precise segmentation of the head part of the sperm cells is required to calculate the length, width, length-to-width ratio and area of the sperm head [19]. However, due to the lack of chemical specificity of QPI technique, precisely locating the boundary between head and mid-piece of the sperm cell is not possible. Secondly, segmentation algorithm and human expertise play an important role in measuring the length, width and area of the head part of the cells. In addition, decoupling of thickness and local refractive index is crucial to determine the correct value of morphological features, such as volume, surface area, surface-to-volume ratio and sphericity [20]. Therefore, it is recommended to use modern DNN architectures which essentially support abstract characterization of all the physical and relevant statistics to these features. Also, it does not require segmentation of different parts and utilization of morphology of all parts to detect the changes in phase value.

8.2.2.2 Modern deep learning approach

Modern deep learning architecture i.e., CNNs support abstract characterization of texture, refractive index profiles, scattering and absorption coefficient distribution as well as relevant statistics to these features. CNN has proved its ability in various biomedical applications such as classification, segmentation and generation of pseudo high-resolution images and so on. A popular approach to using CNN is to perform transfer learning, i.e., use a pretrained CNN on object classification datasets.

The recent application of DNN architectures used for classification of sperm cells is shown in figure 8.4 where a total of 70% of the phase datasets is used for the training while 30% is used for testing different CNN models. Training with sufficiently large datasets is important to overcome overfitting and underfitting in the neural network while validating against the testing datasets. Figure 8.4(a) represents the confusion matrices of the sperm cells testing datasets using different pretrained CNN models. These matrices show the ability of CNN to classify unknown datasets. Rows and columns of the matrix indicate the predicted class and the ground truth, respectively. Diagonal elements of the confusion matrix show the correct predictions of the sperm datasets while off-diagonal elements represent incorrect prediction. For example, a confusion matrix of AlexNet in figure 8.4(a): total 720 phase images of the normal sperm cells are tested where 525 images are predicted correctly and 13, 103 and 79 are the wrong predictions. Similarly, 525, 520 and 637 are the correct prediction for Ethanol, H_2O_2 and cryopreservation, respectively. Performance of the other architecture, i.e., GoogLeNet, VGG-16, VGG-19, Inception-ResNet-V2, ResNet-50 and ResNet-101 can be also seen in figure 8.4. ResNet-101 provides the best accuracy of 85.6% for classifying the sperm phase images.

8.2.3 Phase retrieval and virtual staining in QPI

Recent advancements also show the potential applications of deep learning in recovering phase map of the object. The data driven approach to recover the phase map from a hologram overcomes the disadvantage of conventional computational technique such as phase unwrapping artefact, iterative reconstruction and most importantly fine- tuning of user-defined parameters for convergence of complex

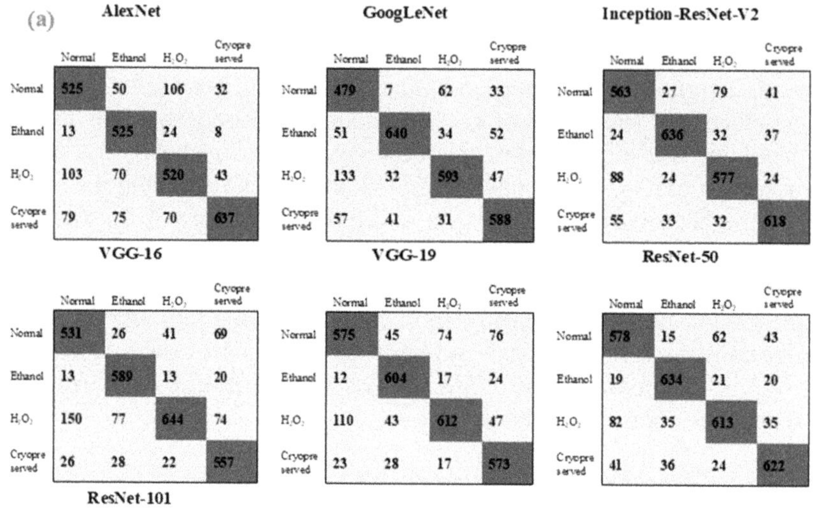

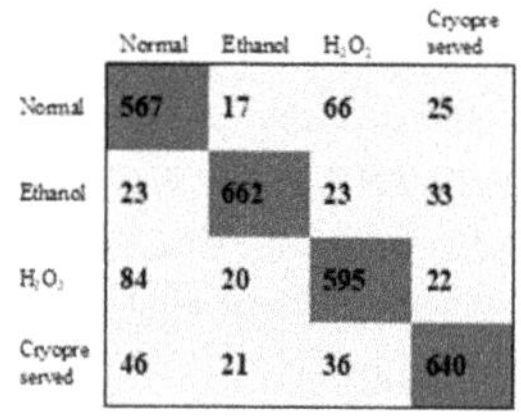

(b) Per class accuracy of ResNet-101

	Sensitivity	Specificity	Accuracy
Normal	84.0	92.5	78.8
Ethanol	89.3	96.9	91.9
H_2O_2	82.5	93.7	82.6
Cryopreserved	86.1	95.8	88.9

Figure 8.4. Testing performance of the CNN on the phase images of normal and treated sperm cells. (a) Confusion matrices of different DNN for classification of four different classes of sperm cells. (b) Per-class sensitivity, specificity, and accuracy of ResNet-101. Adapted with permission from [16], copyright 2020, The Author(s). With permission of Springer.

value image. In addition, lack of chemical specificity is the main bottleneck with phase images, therefore making it difficult to analyze and compute the cellular changes in the object.

To address the aforementioned challenges, deep learning has recently demonstrated recovery of the phase map and virtually staining the phase image using a generative adversarial network (GAN). GAN is a special kind of neural network consisting of two building blocks, i.e., generator and discriminator. The generator (G) learns best mapping from input image and random noise vector to target image during the training process. The discriminator (D) is trained to outperform the generator by computing the cost between mapped image generated by the generator and the corresponding ground truth. In addition, such types of approaches use first the accurate registration between input and target classes. After training, the network can automatically transform the input image into the desired labelled image. An example image of transforming a hologram into fluorescent images is shown in figure 8.5. A trained network takes a reconstructed label-free QPI of a tissue sample as input and then outputs the equivalent histologically stained sample. Therefore, using CNN it is possible to generate pseudo contrast without histological staining of the sample.

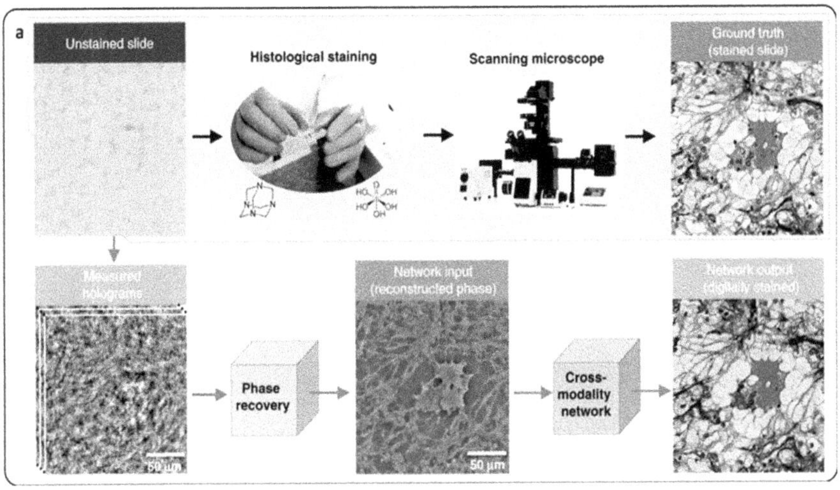

Figure 8.5. (a) Phase staining using DNN [10]. A trained network takes reconstructed phase images of unstained tissue sample as input and generates a histologically stained sample. Adapted with permission from [10], copyright © 2019, The Author(s). With permission of Springer.

8.2.4 High resolution quantitative phase imaging using machine learning

GAN can also be used to predict high resolution phase imaging from low resolution images. The GAN architecture comprises simultaneous training of generator (G) and discriminator (D) [21, 22]. The generator model contains various convolutional layers, as shown in figure 8.6, and uses the input dataset to generate mapped images. On the other hand, a discriminator computes the cost between target ground truth and generated images. While differentiating between ground truth and generated images, the discriminator calculates the losses and gives feedback to the generator for fine-tuning.

Previous work has shown the phase+CNN framework for generating a high resolution phase image, as shown in figure 8.7 [12]. The workflow can be divided into three parts, i.e., data acquisition, image registration and the prediction of a high-resolution image. RBC samples can be prepared by mixing phosphate buffered saline (PBS) solutions. After mixing with the solution, centrifugation of 10–15 min can be done to isolate the RBCs from other components such as plasma and debris. A few drops of the isolated RBC sample was pipetted into a polydimethylsiloxane (PDMS) chamber for the imaging. The PDMS chamber is prepared on top of the reflecting silicon slides since a Linnik-type system is used for the phase imaging.

The comparison of low-resolution images and predicted high resolution phase for RBC datasets is shown in figure 8.8. Figures 8.8(a) and (b) depict the interferogram and reconstructed phase map for 10×, 0.25 NA lens. The regions of interest (ROI) 1 and 2 are upsampled (figures 8.8(c_1) and (f_1)) exactly the same pixel size as HR image. The upsampled ROI is compared with the predicted phase and the experimentally acquired ground truth image. The predicted phase of ROI 1 (figure 8.8(e_1)) and ROI 2 (figure 8.8(h_1)) shows notable resolution enhancement and is matched approximately with the ground truth (figures 8.8(d_1) and (g_1)). Further, spatial frequency analysis can be done to show resolution enhancement in

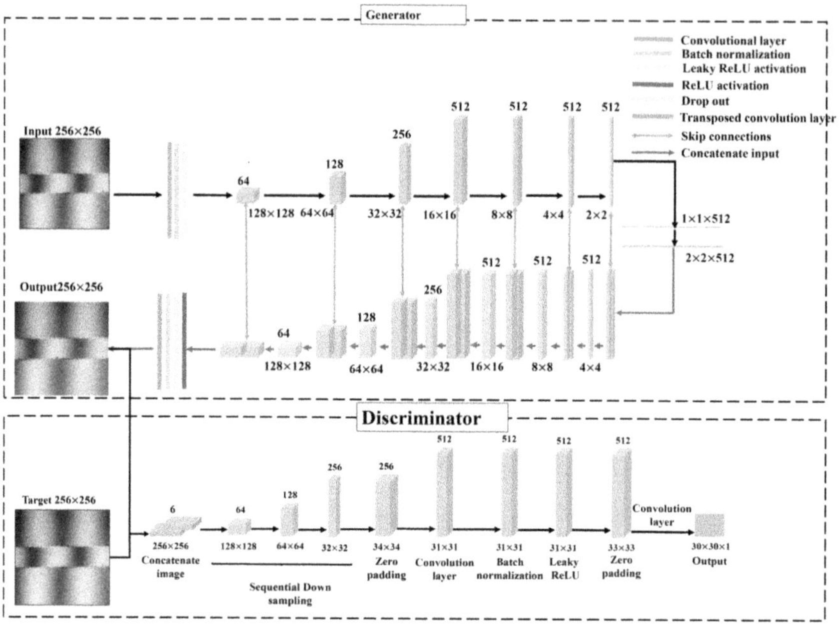

Figure 8.6. GAN architecture for high-resolution phase imaging using phase-shifting interference microscopy. Reproduced with permission from [23], copyright 2021 Wiley-VCH GmbH.

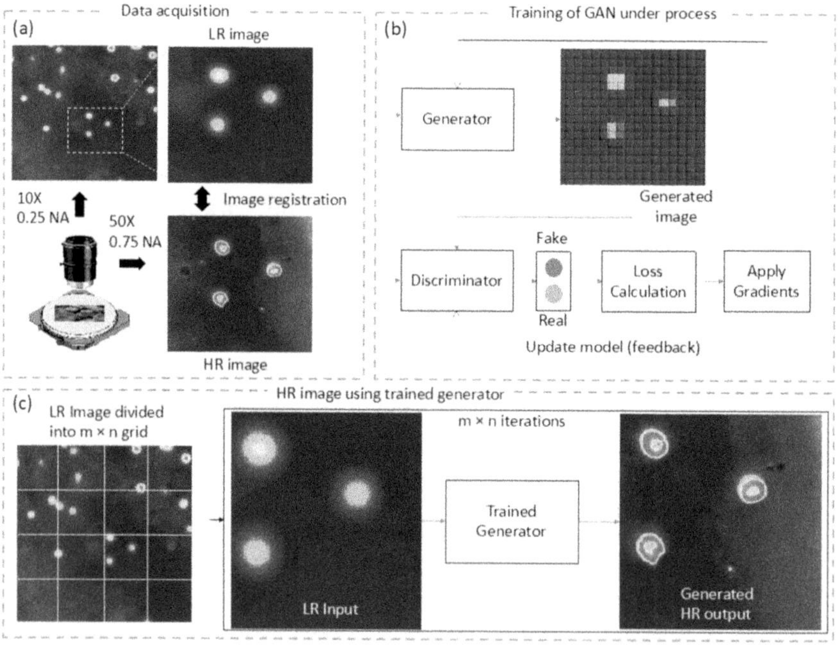

Figure 8.7. High-space bandwidth phase imaging using partially spatially coherent quantitative phase microscopy assisted with GAN: (a) low resolution and high-resolution phase images are acquired by 10×, 0.25 NA and 50×, 0.75 NA, respectively. (b) Training workflow of GAN, (c) HR image prediction using trained generator. Adapted with permission from OSA [12], copyright 2022 Optica Publishing Group. All Rights Reserved.

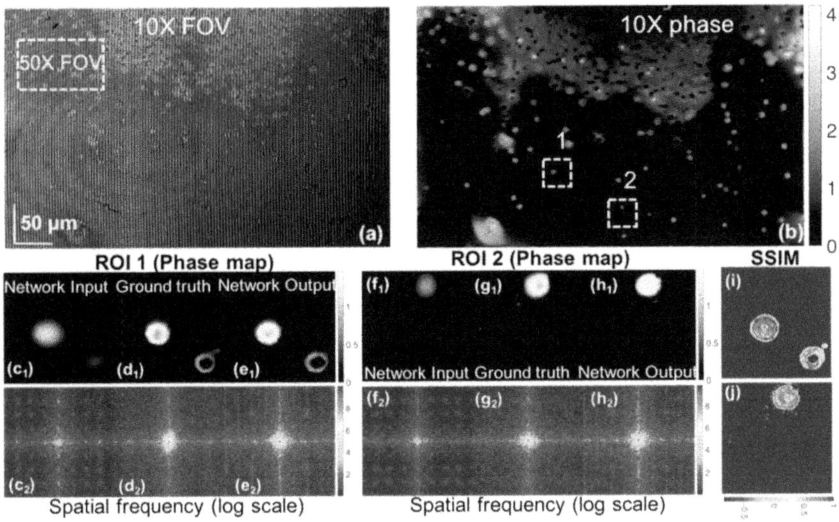

Figure 8.8. High resolution phase prediction for human RBC. (a and b) 10× interferometric and reconstructed phase map. The small ROI 1 and 2 compare with the ground truth and network output. (i) and (j) Structured similarity index (SSIM) between predicted phase and ground truth of ROI 1 and 2, respectively. SSIM for ROI 1 and ROI 2 is found to be 0.96 and 0.97, respectively. Adapted with permission from [12], copyright 2022 Optica Publishing Group. All Rights Reserved.

the prediction. A spatial frequency spectrum for ROI 1 and 2 is shown in figure 8.8(c_2–e_2) and (f_2–h_2), respectively. Broadening in spatial frequency (log scale) spectrum of the predicted high-resolution image is another indication to achieve a higher frequency component using a trained network.

The spatial frequency spectrum of the network predicted images matches closely to the ground truth phase image. Nonetheless, difference in spatial frequency and structural information can be quantified by calculating SSIM index [24]. Figures 8.8(i) and (j) represent SSIM between ground truth and predicted phase of ROI 1 and 2, respectively. The SSIM index varies between −1 and 1. The SSIM value of 1 will be achieved only if the predicted and ground truth images are identical. The SSIM indexes for ROI 1 and ROI 2 are found to be 0.96 and 0.97, respectively. These values show that predicted images and the ground truth phase images match closely. However, some mismatch occurs due to the illumination and phase related artifacts which cannot be neglected completely and hence affect the network prediction and the spatial frequency spectrum of the predicted images.

8.3 Practical prescription for machine learning in QPI

A careful examination must be required to interpret the QPI+DNN framework. For example, in the case of virtual staining, although the phase and fluorescence images are mapped and virtual staining is clearly visible, the fine information of the cells is slightly mismatched due to imperfection of the microscope system, illumination geometry and image registration. Further, in data driven-based high space bandwidth approach, other artefacts due to illumination, phase sensitivity (spatial and temporal), and phase reconstruction + unwrapping algorithm cannot be completely avoidable. Any phase information equal to or smaller than the spatial phase sensitivity, i.e., ±20 mrad in this case likely represents the artefact. Also any mechanical displacement even in the nanometer range might create trouble in precisely matching of HR and predicted phase map. The additional phase unwrapping artefact can create some problems, for instance, Goldstein *et al* [25] and minimum Lp norm method [26] generated streaky artefacts in the unwrapped phase, especially in the case of a thick sample. Furthermore, the sample preparations and sub-pixel matching of phase and stained sample pairs increases the challenges to predict a virtual stained image.

8.4 Outlook

QPI+DNN approach can be used for various applications such as automated classification between normal and abnormal cells. QPI is a promising technique which provides high contrast quantitative information of the sample in a label-free manner and therefore has an edge over the conventional intensity-based techniques. A partially spatially coherent based QPM system can be used to extract the phase value of the thinnest part of the cells with high spatial phase sensitivity (± 20 mrad). A variety of DNN architectures provide good classification accuracy to separate the 2–6 types of cells into diagnostically relevant classes. In future, fully automated classification of healthy sperm cells as per World Health Organization (WHO) criteria could be an intermediate tool for the expert. QPI+DNN framework could be very useful, especially for real-time selection of healthy and progressive sperm cells that can be used for improving the success of fertilization.

On the other hand, deep learning-based phase recovery, virtual staining and resolution enhancement have broad applications in biomedical imaging, data interpretation, material science, among others. These data driven approaches can be useful to improve robustness, computational artefacts, chemical specificity and space-bandwidth product in both coherent and incoherent phase imaging. In the first site, where QPI offers morphological changes and offers quantitative information of the sample, deep learning provides an automated platform to analyze, segment and classify the datasets into their diagnostically relevant classes. In summary, integrating QPI+AI offers a powerful data driven approach for statistical analysis and unique tool to analyze the phase images. It is also important to understand the pitfalls of AI in biomedical imaging to understand the potential benefits of data driven learning.

References

[1] Kermany D S *et al* 2018 Identifying medical diagnoses and treatable diseases by image-based deep learning *Cell* **172** 1122–31

[2] Ran A R *et al* 2021 Deep learning in glaucoma with optical coherence tomography: a review *Eye* **35** 188–201

[3] Butola A *et al* 2020 Deep learning architecture 'LightOCT' for diagnostic decision support using optical coherence tomography images of biological samples *Biomed. Opt. Express* **11** 5017–31

[4] Butola A *et al* 2019 Volumetric analysis of breast cancer tissues using machine learning and swept-source optical coherence tomography *Appl. Opt.* **58** A135–41
[5] Joshi D *et al* 2021 Label-free non-invasive classification of rice seeds using optical coherence tomography assisted with deep neural network *Opt. Laser Technol.* **137** 106861
[6] Gan Y, Tsay D, Amir S B, Marboe C C and Hendon C P 2016 Automated classification of optical coherence tomography images of human atrial tissue *J. Biomed. Opt.* **21** 101407
[7] Devalla S K *et al* 2019 A deep learning approach to denoise optical coherence tomography images of the optic nerve head *Sci. Rep.* **9** 1–13
[8] Athanasiou L S *et al* 2014 Methodology for fully automated segmentation and plaque characterization in intracoronary optical coherence tomography images *J. Biomed. Opt.* **19** 026009
[9] Jo Y *et al* 2017 Holographic deep learning for rapid optical screening of anthrax spores *Sci. Adv.* **3** e1700606
[10] Rivenson Y *et al* 2019 PhaseStain: the digital staining of label-free quantitative phase microscopy images using deep learning *Light: Sci. Appl.* **8** 1–11
[11] Rivenson Y, Wu Y and Ozcan A 2019 Deep learning in holography and coherent imaging *Light: Sci. Appl.* **8** 1–8
[12] Butola A *et al* 2020 High space-bandwidth in quantitative phase imaging using partially spatially coherent digital holographic microscopy and a deep neural network *Opt. Express* **28** 36229–44
[13] Moon I and Javidi B 2007 Three-dimensional identification of stem cells by computational holographic imaging *J. R. Soc. Interface* **4** 305–13
[14] Javidi B, Moon I, Yeom S and Carapezza E 2005 Three-dimensional imaging and recognition of microorganism using single-exposure on-line (SEOL) digital holography *Opt. Express* **13** 4492–506
[15] Go T, Kim J H, Byeon H and Lee S J 2018 Machine learning-based in-line holographic sensing of unstained malaria-infected red blood cells *J. Biophoton.* **11** e201800101
[16] Butola A *et al* 2020 High spatially sensitive quantitative phase imaging assisted with deep neural network for classification of human spermatozoa under stressed condition *Sci. Rep.* **10** 1–12
[17] Dubey V *et al* 2019 Partially spatially coherent digital holographic microscopy and machine learning for quantitative analysis of human spermatozoa under oxidative stress condition *Sci. Rep.* **9** 1–10
[18] Singh R *et al* 2020 Quantitative assessment of morphology and sub-cellular changes in macrophages and trophoblasts during inflammation *Biomed. Opt. Express* **11** 3733–52
[19] Mirsky S K, Barnea I, Levi M, Greenspan H and Shaked N T 2017 Automated analysis of individual sperm cells using stain-free interferometric phase microscopy and machine learning *Cytom. Part* A **91** 893–900
[20] Girshovitz P and Shaked N T 2012 Generalized cell morphological parameters based on interferometric phase microscopy and their application to cell life cycle characterization *Biomed. Opt. Express* **3** 1757–73
[21] Goodfellow I *et al* 2014 Generative adversarial nets *Advances in Neural Information Processing Systems* (Cambridge, MA: MIT Press) pp 2672–80
[22] Pathak D, Krahenbuhl P, Donahue J, Darrell T and Efros A A 2016 Context encoders: feature learning by inpainting *Proc. IEEE Conf. on Computer Vision and Pattern Recognition* 2536–44

[23] Bhatt S, Butola A, Kanade S R, Kumar A and Mehta D S 2021 High-resolution single-shot phase-shifting interference microscopy using deep neural network for quantitative phase imaging of biological samples *J. Biophoton.* **14** e202000473
[24] Wang Z, Bovik A C, Sheikh H R and Simoncelli E P 2004 Image quality assessment: from error visibility to structural similarity *IEEE Trans. Image Process.* **13** 600–12
[25] Goldstein R M, Zebker H A and Werner C L 1988 Satellite radar interferometry: two-dimensional phase unwrapping *Radio Sci.* **23** 713–20
[26] Ghiglia D C and Romero L A 1996 Minimum L^p-norm two-dimensional phase unwrapping *J. Opt. Soc. Am.* A **13** 1999–2013

Chapter 9

Multi-spectral and hyper-spectral phase microscopy

Recently, multi-spectral and hyper-spectral imaging has been rigorously used to obtain information about the object and sample for various spectral components. These technologies are mainly used in drone photography and videography for agricultural fields and landscapes, providing rich information about the objects. More recently, these technologies have also been used for microscopy of biological samples. In this chapter we describe various optical configurations for multi-spectral and hyper-spectral phase microscopy. A variety of spectral filters and wavelength-scanning systems are also described. Algorithms for image reconstruction, and experimental results of multi-spectral and hyper-spectral phase microscopy of red blood cells (RBCs) are presented. These QPM technologies have a great future for quantitative imaging of biological samples.

9.1 Introduction

The optical properties of a biological specimen greatly determine the usage of light for diagnostic as well as for therapeutic purposes. Diagnostic applications mainly depend on the ability of light to enter the tissue, probe it and then exit it to reveal the structure of tissue while the therapeutic application depends on the ability of light to infiltrate the tissue and accumulate the energy depending on the absorption property of the tissue. This makes the determination of optical properties of the tissue an important undertaking even for device development and diagnosis of various diseases. The optical properties of tissue are mainly defined in terms of its absorption coefficient, scattering coefficient, scattering structure function and the refractive index [1]. All these properties are wavelength dependent and their knowledge at various wavelengths becomes essential. Detailed information about the dependence of these optical properties (i.e. absorption and scattering) on the wavelength of light variation has been described in [1, 2]. To estimate their dependency on wavelength of light, the data/information must be acquired in the spectral domain as well.

doi:10.1088/978-0-7503-3987-2ch9

Acquiring spectral information about a specimen is usually done in three dimensions, that is: two are the spatial dimensions (x, y) and the third is the spectral dimension (λ). The dataset is generated usually in the form of a three-dimensional data cube. The three-dimensional data cube can be easily visualized in figure 9.1. It shows the data cube having (x, y, λ) as the three axes of the data cube.

Spectral imaging is mainly divided into two subsets, depending on the number of spectral bands information captured, and the continuity of the spectral bands employed. The two subsets comprise multi-spectral and hyper-spectral imaging. These can be differentiated based on mainly two factors. Firstly, multi-spectral imaging mostly refers to capturing a smaller number of spectral bands usually up to 10 spectral bands while hyper-spectral imaging captures 10 to 100 of spectral bands. Hyper-spectral imaging represents continuously measuring the spectrum while multi-spectral imaging has non-continuous spectral bands [2] which may be uniformly or non-uniformly distributed. The multi-spectral 2D interferograms can be represented by the expressions $I_i(x, y, \lambda_i) = I_0(x, y, \lambda_i)[1 + V_i\gamma(z)\cos\{\phi_i(x, y, \lambda_i) - \delta\}]$, where I stands for various colour wavelengths from deep red, red, amber, green, blue and violet. $I_0(x, y, \lambda_i)$ is the background intensity, V_i is the fringe contrast and $\gamma(z)$ is

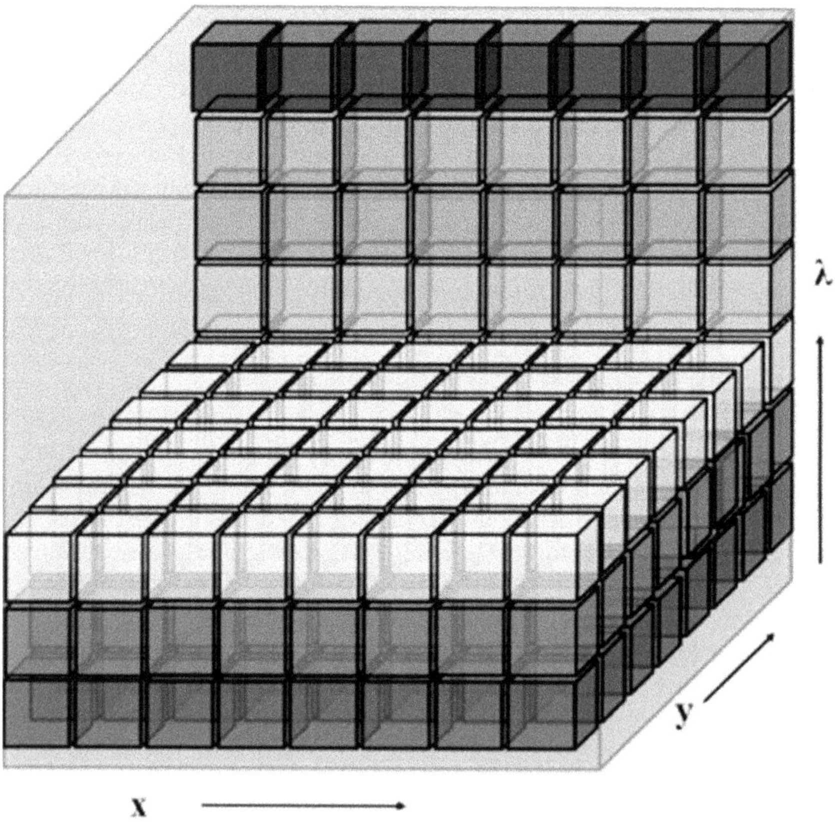

Figure 9.1. The three-dimensional multi-spectral/hyper-spectral data cube having x, y and λ as the three axes [2].

coherence envelop of the light source and defined as $\gamma(z) = \exp[-2\pi(z - z_0/l_c)^2]$ with coherence length l_c and δ is the incremental phase shift to be introduced for phase-shifting interferometry to reconstruct a phase map $\phi_i(x, y, \lambda_i)$.

9.2 Multi-spectral phase microscopy

Multi-spectral, in general, refers to the term of using multiple tuneable wavelengths as the source and the information is recorded at all these multiple wavelengths. In multi-spectral phase microscopy, a series of interferograms of the specimen are recorded at multiple wavelengths in order to get their spectral information, phase information or their absorption/transmission properties at various wavelengths [3–5]. Multi-spectral phase microscopy is widely used in the field of biomedicine to quantitatively measure the various morphological, chemical and optical properties of the specimen used [6–14]. Figure 9.2(a) shows the schematic diagram of a multi-spectral phase microscope. It comprises a white light source; the incoming light is collimated with a collimating lens L_1 and passes through a linear variable filter (LVF) after passing through an aperture stop A. The light from the filter LVF is made to fall on the specimen which is further collected by the microscope objective (MO). The information collected by the microscopic objective is relayed by the tube lens and is further incident on the interferometric unit for phase measurement and the final image of the sample is obtained by placing a monochrome camera (CCD) on the imaging plane using lens L_2. Finally, the multi-spectral cube can be generated by further analysis. This interferometric unit could be a common-path or nearly-common-path system such as lateral shearing interferometer, lateral shear plate, Fresnel bi-prism or diffraction phase microscope. Figure 9.2(b) shows the broad-band spectrum of a white light source which passes through the LVF mounted on a linear motorised stage.

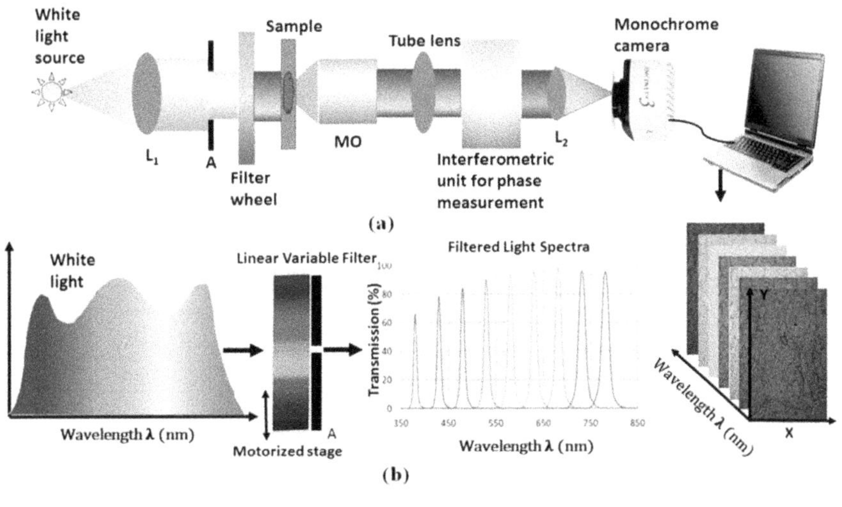

Figure 9.2. (a) Schematic diagram illustrating the multi-spectral phase microscope; L_1 and L_2 are lenses, A—aperture, MO—microscope objective lens. (b) Spectrum of white light source followed by linear variable filter, A—aperture, filtered spectra and multi-spectral frames.

An aperture A is placed in front of the LVF. By means of translating the linear variable filter, as shown in figure 9.2(b), a narrow band of single-color light is transmitted through the aperture and the entire broad-band spectrum is scanned. The output of this system is shown in figure 9.2(c) as the spectral profile of the filtered light. Figures 9.3(a)–(c) show the various combinations of narrow-band filters to scan or filter the broad-band light, figure 9.3(a) is the most generally used device which is a colour filter wheel, these colour filters (VIBGYOR) are individually fabricated using thin film technology, figure 9.3(b) is a linear variable filter in which a tapered cavity with varying thickness is inserted between two dielectric mirror structures and all these layers are coated on a substrate, as shown in figure 9.3(b), due to the variation in thickness different colours are transmitted from different locations. The working principle of this device is similar to Fabry–Perot interferometer with multiple beam interference, figure 9.3(c) shows a similar device, as shown in figure 9.2(b), but with a step-like thickness variation. All these combinations can be used as spectral filtering for multi-spectral quantitative phase microscopy. With the current advancements in the solid-state devices, we can also use a combination of multi-colour light-emitting diodes (LEDs) as shown in figure 9.4(a) which is quite cheap. LEDs with all the colours in the visible range also in the near IR range are available in the market, a linear array of multi-colour LEDs is mounted on a single printed circuit board (PCB), as shown in figure 9.4(a). This can be controlled by a controller unit connected with a power supply. By means of linear translation, a single-colour LED is switched ON sequentially, and by translating the LED array using a motorised stage we can have the light output as shown in figure 9.4(b). In place of white light source with filter wheel combination, a simple multi-colour LED array can be used as a light source.

In multi-spectral phase microscopy in many situations, we require a highly monochromatic and long temporal coherence light source. Currently, laser diodes of different colours are available on the market. These are compact, relatively low cost, high brightness and rugged. By means of combining the multiple colour laser diode, as shown in figure 9.4(a), we can develop a multiple colour quantitative phase

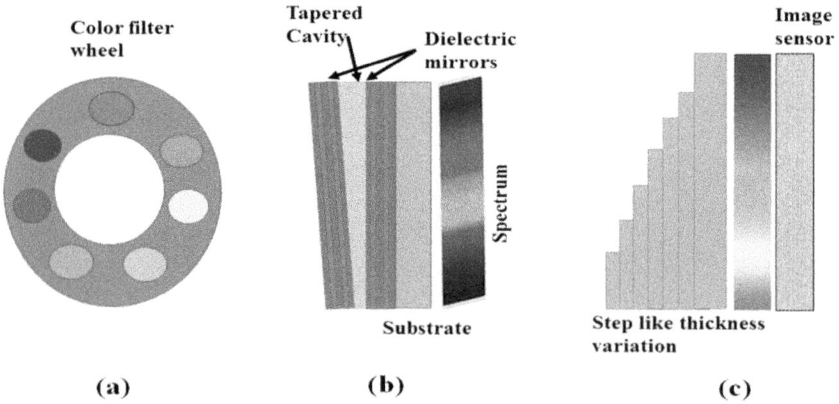

Figure 9.3. Schematic diagram depicting a variety of wavelength tunable filters; (a) colour filter wheel, (b) tapered cavity continuous variable tunable filter and step-like variable thickness tunable filter.

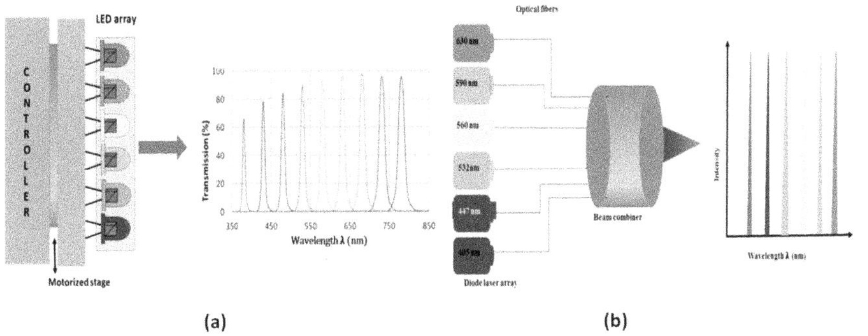

Figure 9.4. (a) Multi-colour LED with controller and the output spectra, (b) multi-colour laser diodes and the corresponding spectra.

microscope. Figure 9.4(a) again shows the different diode laser coupled with a fibre and combined with a beam combiner to have a spectral output, as shown in figure 9.4(b), by switching on the laser diode sequentially. This multi-spectral diode laser system can be used in place of white light source in figure 9.2(a). We can replace the white light source with an LED array or a diode laser array ranging from 405 nm to 620 nm (VBGYOR).

9.3 Hyper-spectral quantitative phase microscopy

Hyper-spectral imaging (I) is slightly different from multi-spectral imaging. In hyper-spectral imaging the scene/object is illuminated using a broad-band light source and spectral filtering is done at the detector end by means of inserting the continuous variable colour filter or a dispersing element. In hyper-spectral imaging technology one records data in hundreds of wavelengths thereby making it useful in differentiating various chemical samples even when they have overlapping emission and reflectance spectra. Hyper-spectral imaging finds its application in biomedicine, which involves development of biosensors and quantum dots and organic fluorophores. Figure 9.5(a) shows a typical hyper-spectral phase microscopic system. The incoming light from white light source is collimated with the help of collimating lens (L_1) and passes through an aperture stop A. The light from the aperture is made to fall on the specimen which is further collected by the MO. The information gathered by the MO is transmitted through the tube lens and is further incident on the dispersive element and the final data is obtained by placing a colour CCD camera on the imaging plane using lens L_2. Finally, the hyper-spectral cube can be generated by further analysis. For dispersive element we can use a prism, grating element or a Fabry–Perot etalon, as shown in figure 9.5(b), which provides the different wavelengths onto the different positions.

9.4 Optical configurations of multi-spectral and hyper-spectral phase microscopy

Quantitative phase imaging (QPI) or phase microscopy is one of the most essential techniques which have been utilised in the field of modern healthcare. It helps in

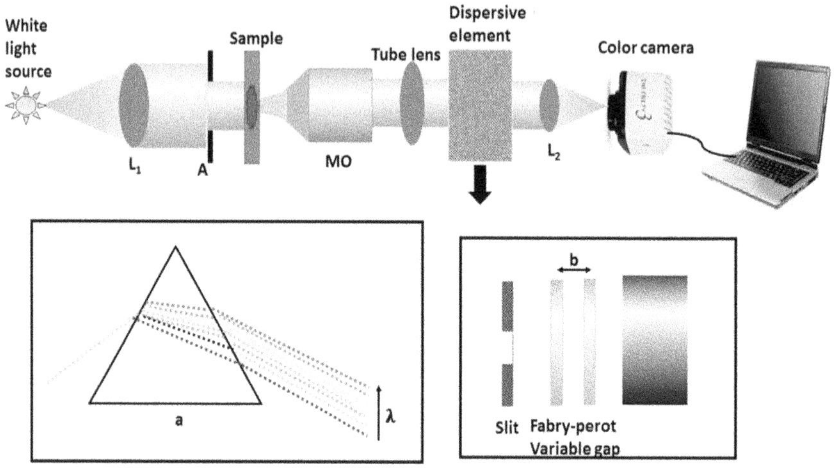

Figure 9.5. (a) Schematic diagram of a hyper-spectral microscopic system, L_1 and L_2 are lenses, A—aperture, MO—microscope objective lens; panel (b) shows the dispersing element for continuous spectral filtering unit; left a prism-based and right Fabry–Perot interferometer with variable gap.

getting quantitative information about the sample with high sensitivity and without the need of tagging. Phase microscope helps us in getting phase information about the specimen which directly relates to refractive index of the sample and thus, assists us in obtaining various optical parameters at various wavelengths. These optical parameters are wavelength dependent, and their spectral information is vital for healthcare diagnosis. Multi-spectral/multi-wavelength phase microscopy has been achieved using various methods and techniques including multiple light sources. It can be realized using broad-band light source connected with a filtering device, or using multiple light sources fabricated into a single module such as multiple LEDs. In figure 9.6, the multi-spectral phase microscopy has been implemented using supercontinuum source coupled with an acousto-optic tuneable filter (AOTF) (YSL SC-PRO 7 and YSL AOTF-PRO). The multiple wavelengths are produced with the help of AOTF functionality. The light from the source is collimated and made to pass through the sample. The light scattered through the sample is collected by the objective lens and then it passes through the Fresnel bi-prism. The bi-prism splits the beam into two beams that are interfered on the camera plane. The lenses after the bi-prism are used for the magnification such that sufficient fringe density is obtained on the camera plane, which satisfies Nyquist criteria. The color CCD camera is employed to capture the multi-spectral interferograms with the change in wavelength at the source end.

Furthermore, multi-spectral digital holography can be implemented for label-free QPI that utilizes the commercial microscope enabled supercontinuum light source. The optical configuration, as shown in figure 9.7, comprises an inverted microscope coupled with multiwavelength digital holographic microscope and the source is a tuneable laser light source combined with AOTF. Digital holography microscopy is achieved with the help of a Michelson interferometer and holograms are recorded at different wavelengths from 470 nm to 700 nm and 800 nm to 850 nm [15].

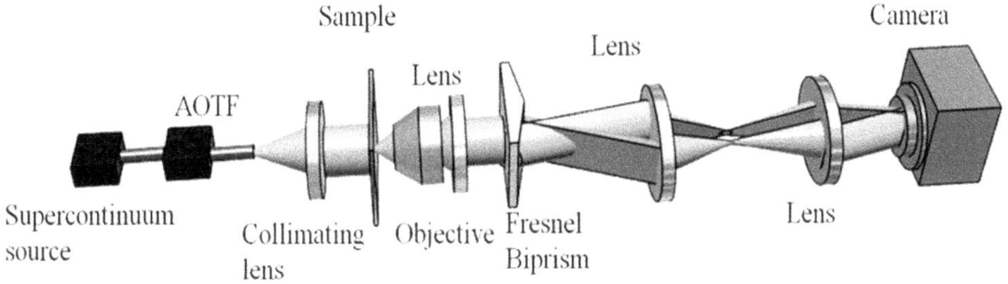

Figure 9.6. Experimental configuration of multi-spectral phase microscopy using supercontinuum light source coupled with an AOTF filter.

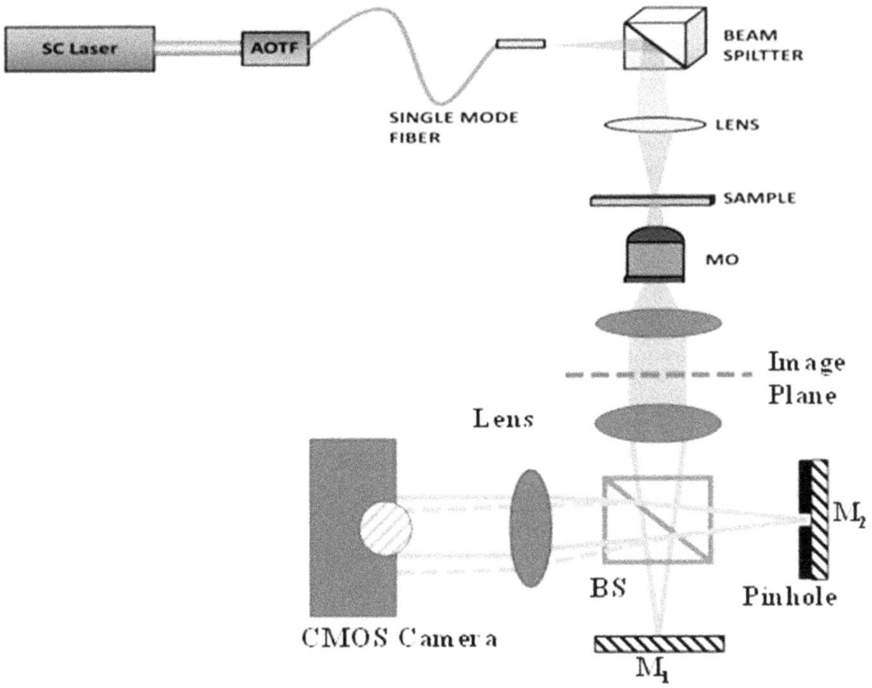

Figure 9.7. Optical configuration of multi-spectral digital holography in reference. Adapted from [15], copyright 2022 SPIE.

In addition to using broad-band light source in combination with tunable filter, various multiple wavelength light sources may also be used. The authors [5] have carried out multi-spectral quantitative phase microscopy using multi-coloured LEDs. As LEDs are low cost and easy to handle, multiple wavelength LEDs are mounted together on a single unit. The light from the LEDs is coupled to a quantitative phase microscope. Their system involved the use of a Mirau interferometric objective lens which divides the incoming light into beams in a 50:50 ratio utilizing a beam splitter. One beam behaves as reference beam as it travels towards the in-built reference mirror while the other beam acts as object beam as it travels towards the sample. The object beam and the reference beam interfere to produce

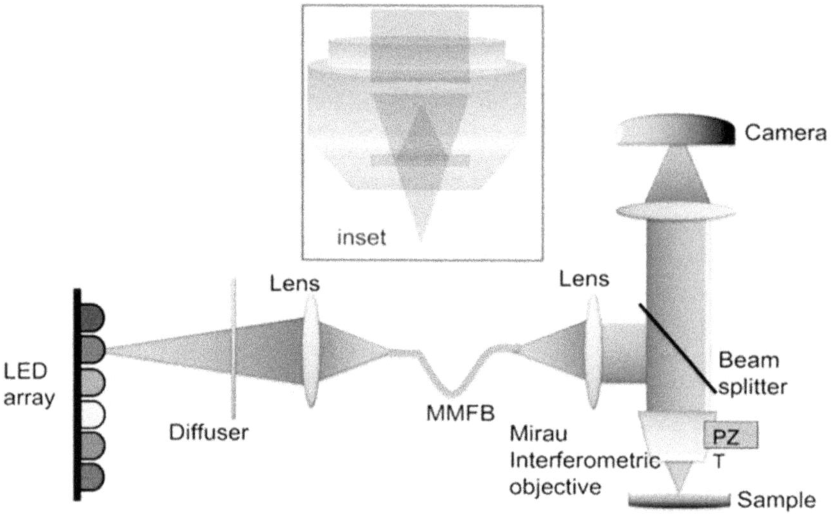

Figure 9.8. Experimental set-up of Mirau-based multi-spectral quantitative phase microscopy. (inset) Mirau type objective lens. Adapted with permission from [5], copyright 2016 Optical Society of America.

the interference pattern. Their objective was attached to a piezo-electric transducer (PZT) translational stage which provided phase-shifted interferograms. The experimental set-up and objective lens (Mirau) is shown in figure 9.8.

In reference [6] the authors have used a wide-band light source for the imaging purpose. The light from the wide-band source is coupled to a transmission light microscope which is further connected to a compact off-axis Tau (τ) interferometer. The multi-spectral imaging is achieved with the help of an acousto-optic tuneable imaging filter and a digital camera. Their wide-band source consisted of a super-luminescent diode Superlum (S-785-B-I-15 ($\Delta\lambda$ = 770–810 nm)) connected to a single-mode fibre. They utilized the single-mode fibre to increase the spatial coherence of the source. The block diagram of their optical configuration is shown in figure 9.9, which consists of a wide-band light source coupled to a transmission microscope further coupled to a τ-interferometer and AOTF and camera. In this configuration the wavelength slicing has been carried out at the detector end of the experiment set-up instead of the source end as described in the previous cases.

In reference [16] the authors have demonstrated digital holographic phase microscopy based on swept source. They achieved wavelength scanning with the help of swept source that is Superlum BS-840-1, 802–878 nm which was filtered spatially and thereby collimated. The light which was collimated was then coupled to a digital holographic phase microscope based on a Mach–Zehnder-type interferometer.

In reference [17] the authors carried out spectrally resolved white light interferometry for acquiring multi-wavelength information. They replaced the white light source with RGB LEDs and placed a prism-based spectrometer at the output of the white light interferometer. They selected the incoming light with the help of an entrance slit which limited their field of view and the entrance slit is scanned in order to get the complete field of view.

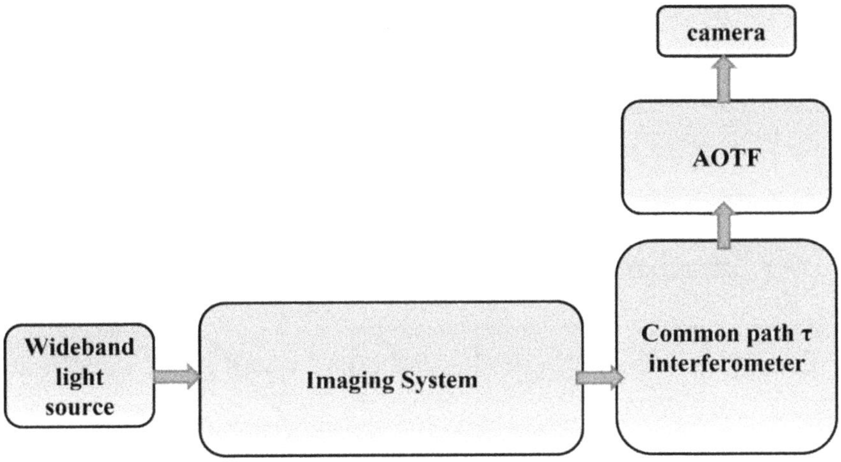

Figure 9.9. Optical configuration of multi-spectral phase microscopy using a τ-interferometer and AOTF and camera.

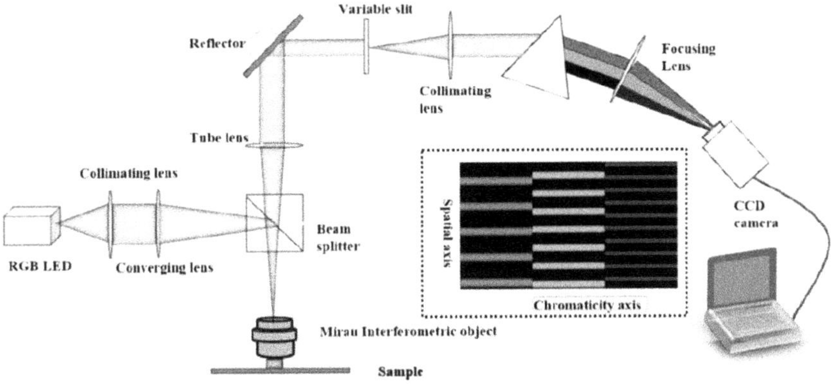

Figure 9.10. Schematic diagram of experimental set-up of SRWLI, in inset simulated spectrally resolved RGB LED interferogram. Adapted from [17], copyright 2017 SPIE.

The experimental set-up of spectrally resolved white light interferometry (SRWLI) is demonstrated in figure 9.10. The figure demonstrates the output of the light microscope is coupled to a prism-based spectrometer and the data is collected using a CCD camera.

The above section described in detail the various optical configurations which have been utilized to carry out multi-spectral phase microscopy lately. As mentioned earlier hyper-spectral phase microscopy is similar to multi spectral phase microscopy in which a greater number of spectral bands are recorded and measurement is carried out in a continuous manner. The systems describing hyper-spectral phase microscopy have been discussed below.

In reference [18] the authors carried out hyper-spectral amplitude and phase imaging of thick and scattering samples using Fourier domain spectroscopy. The images were captured over a wide range of 450 nm to 1700 nm. The authors combined Fourier transform spectroscopy (FTS) with oblique back illumination microscopy to generate

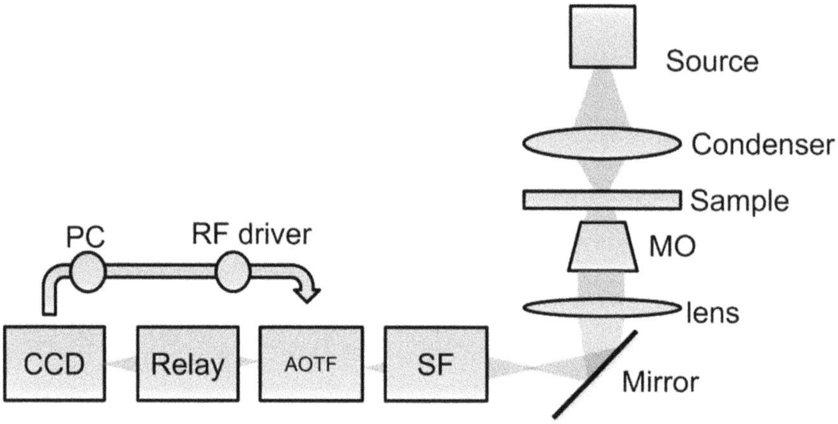

Figure 9.11. A schematic diagram of the experimental set-up in reference. Adapted from [19], copyright 2017 SPIE.

hyper-spectral amplitude and phase images. They demonstrated the use of super-continuum source coupled with motorized filter wheel to generate the wavelength range.

In reference [19] the authors have designed a custom confocal acousto-optic hyper-spectral imaging system that uses an inverted optical microscope (Nikon, Ti–E) and images are captured with the help of a colour CCD camera (Nikon, DS-Fi2). They used an amplitude mask in the Fourier plane for spatial filtering and AOTF was used as a tuneable monochromator. Their system operated in two modes: conventional hyper-spectral imaging data acquisition mode and bandpass special filtering of narrow-band diffracted light mode. Their optical configuration is presented in figure 9.11.

9.5 Light sources for multi-spectral and hyper-spectral phase microscopy

As described in section 9.2, various optical configurations can be used to achieve multi-spectral and hyper-spectral phase microscopy. The configurations can be implemented using various light sources including multiple wavelength lasers, tune-able filters, colour filter wheels, broad-band light source in conjunction with AOTF. The most common include the use of broad-band supercontinuum light source coupled with AOTF. The optical configuration based on supercontinuum light source and AOTF have been described in the previous section. The supercontinuum light source uses a compact laser light source which emits broad-band laser light which has been generated with the help of nonlinear optical phenomena [15]. The nonlinear optical phenomena are usually induced with the help of an ultrashort pulse laser. The supercontinuum source combines properties of laser that are high directionality and high brightness with a broad-band spectrum over a halogen lamp. The supercontin-uum light source finds its application in optical coherence tomography, Fourier transform infrared spectroscopy, infrared spectroscopy, gas measurement analysis etc. The supercontinuum light source in combination with AOTF provides a versatile control over the spectrum used. An AOTF is an electro-optic device which helps us to control the intensity and wavelength of several laser lines. It is based on the use of birefringent crystal whose properties vary on the application of acousto-optic wave. When used in the field of microscopy an AOTF usually consists of trillium dioxide or quartz anisotropic crystal. A PZT is attached to this crystal, such that on the usage of an oscillating radio frequency electrical signal the PZT produces an acoustic wave which transmits into the anisotropic crystal. This propagating wave creates a refractive index redistribution in the crystal making it to act like transmission diffraction grating, which diffracts the incoming laser light into a first order beam. Thus, by adjusting the frequency of the RF signal the period of refractive index variation can be altered. With a change in period of refractive index variation there occurs a change in the wavelength of light diffracted [20].

As shown in figure 9.6, an AOTF when used in combination with a broad-band light source helps in achieving multiple wavelength tunability. The various wavelengths produced with the help of the AOTF are shown in figure 9.12, which were recorded with the help of an LMS 6000 spectrometer having a spectral resolution of ±0.2 nm.

The coherence length of each of the wavelengths can be calculated using the formula given below where λ_0 is the central wavelength and $\Delta\lambda$ is the full width half maximum of the source. Table 9.1 shows that the coherence length ranges from 20.8 μm to 31.0 μm for the spectral bands between 510 and 650 nm.

$$l_c = \frac{2ln_2}{\pi} \times \left(\frac{\lambda_0{}^2}{\Delta\lambda}\right) \tag{9.1}$$

Multi-spectral digital holography microscopy can also be implemented by using a tuneable laser light source (SuperK EXTREME EXR-15 or EXR-9, NKT Photonics A/S, Birkerød, Denmark) accompanied with an AOTF (Super Select 4xVIS/IR or nIR1, NKT Photonics A/S, Birkerød, Denmark). In reference [15] the authors have used the above-mentioned source in combination with an inverted

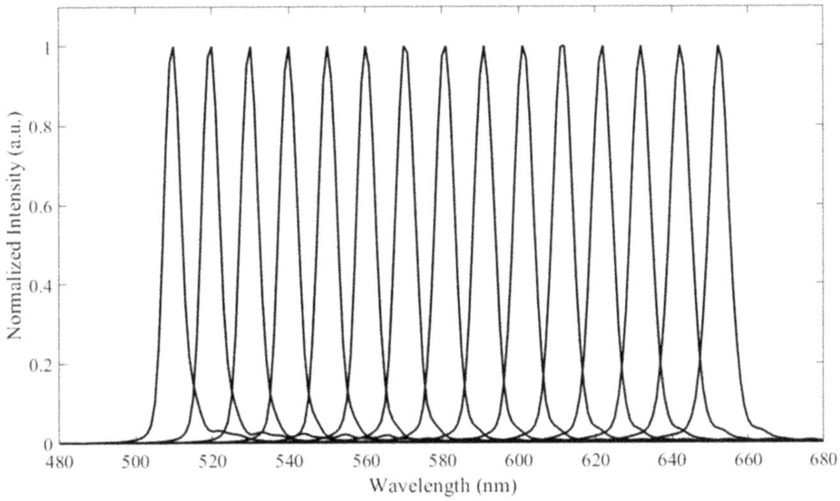

Figure 9.12. Spectrum recorded using supercontinuum source and AOTF.

Table 9.1. Coherence length of light from supercontinuum source coupled with AOTF having central wavelength from 510 nm to 650 nm

Central wavelength (nm)	Coherence length (μm)
510	20.86
520	21.69
530	22.54
540	23.39
550	26.69
560	25.16
570	26.06
580	26.989
590	25.6
600	28.88
610	27.366
620	28.27
630	29.19
640	27.81
650	31.07

microscope and a Michelson interferometer. They recorded holograms from 470 nm to 700 nm or from 800 nm to 850 nm. Their optical set-up is shown in figure 9.7.

In reference [5] the authors have designed a multi-spectral light source by integrating various commercially available low-cost LEDs having different wavelengths. The LEDs were mounted on a PCB board and connected to the power supply individually. The authors placed a diffuser ahead of the low cost LEDs to decrease the spatial coherence. A condenser lens was employed to gather light from the source and it was coupled to a multimode multiple fibre bundle. They employed wavelengths 410 nm, 460 nm, 520 nm, 590 nm, 640 nm and 720 nm to generate the multi-spectral source. The normalized spectra of the LED source are shown in figure 9.13, which have been recorded with the help of Avantes, Avaspec-ULS3648 having 300 lines/mm grating and resolution 0.05 nm. The figure shows that the recorded spectra are nearly Gaussian and their coherence length has been calculated using equation (9.1).

In addition to low cost LEDs, other sources are also used in multi-spectral phase microscopy. In reference [16] the authors have employed a swept source to achieve wavelength scanning and obtained a series of interferograms to achieve accurate phase using spectral domain interferometry.

9.6 Recording devices for multi-spectral and hyper-spectral phase microscopy

Data acquisition in multi-spectral phase microscopy is carried out using a digital camera and AOTF for wavelength slicing. If the wavelength slicing has been carried

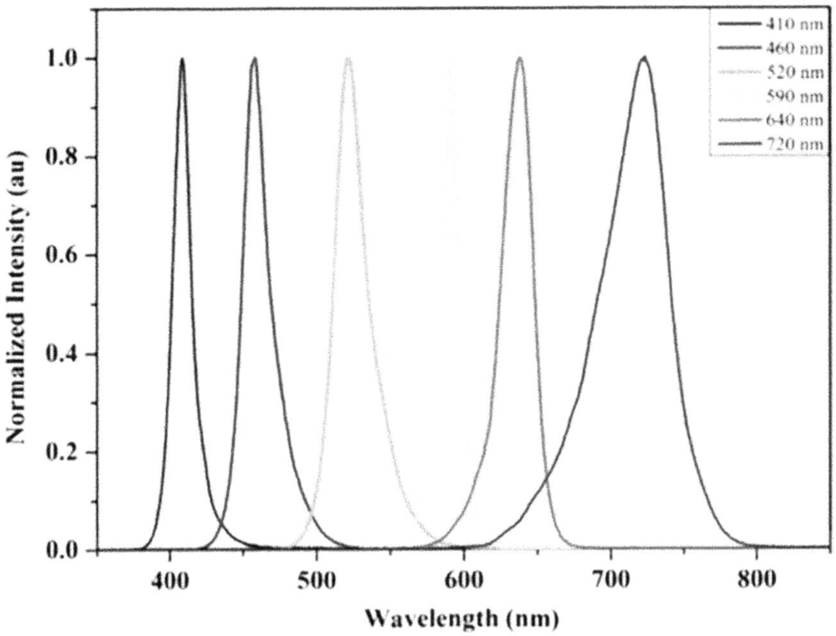

Figure 9.13. Normalized spectra of the LEDs used in multi-spectral quantitative phase microscopy system. Adapted from [5], copyright 2016 Optical Society of America

out at the source end a simple monochrome or colour CCD camera is sufficient to record the interferograms at the various wavelengths. Figure 9.14 shows the interferograms recorded at wavelengths 510 to 650 nm using the set-up displayed in figure 9.6. The variation in wavelength was achieved with the help of LABVIEW code and the corresponding interferograms were recorded using a colour CCD camera (INFINITY 2). The system as shown in figure 9.6 is free from movement of mechanical parts making it highly stable and accurate for multi-spectral diagnosis. In figure 9.14 the upper line shows the recorded interferograms and the lower line displays the zoomed view of the recorded image. The interference fringes can be easily observed in the lower row.

Further, the authors in reference [6] have also used an AOTF attached to a monochrome CCD camera. They have attached the AOTF at the output of a common-path τ interferometer such that the interfering beams coming out of the common-path interferometer are spectral filtered with the help of AOTF. Their tuning range was from 700 nm to 1000 nm with a bandwidth of 4 nm.

Hyper-spectral data acquisition can be carried out using different strategies. The first method is referred to as point-scanning spectroscopy which usually consists of a linear array of detectors which measures the spectral information and then scanning is carried out at different spatial locations to obtain the 3D spectral data cube. The second method of hyper-spectral data acquisition involves recording the information using a 2D detector array and the scanning is carried out in one spatial direction to obtain the data cube. This method is mostly employed in hyper-spectral line-scanning

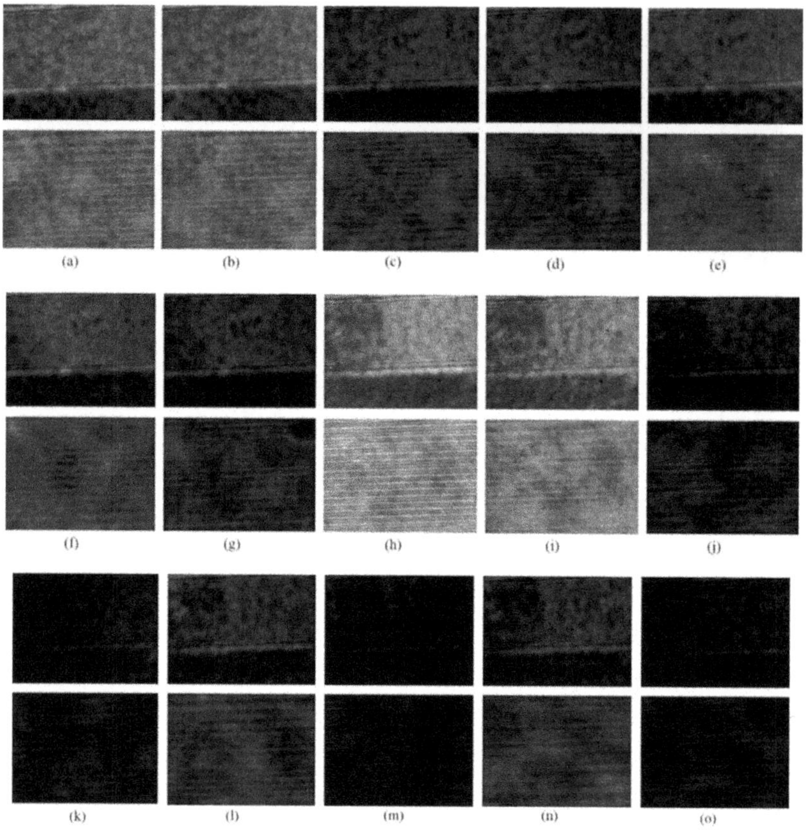

Figure 9.14. The interferograms recorded at various wavelengths using colour CCD camera. (a–o) Interferograms at wavelengths 510, 520, 530, 540, 550, 560, 570, 580, 590, 600 610, 620, 630, 640, 650 nm.

microscopy. The third method used in hyper-spectral imaging may be referred to as wavelength-scanning method in which data is captured in the (X, Y) domain and the scanning is carried out at various wavelengths to generate the 3D data cube. Figure 9.15 shows the data cube where X and Y are the spatial dimension and λ is the spectral dimension; the figure depicts the point-scanning, line-scanning and wavelength-scanning approaches for the data collection. The fourth method includes recording the entire 3D data cube in one shot which leads to enhancement in light throughput in comparison to the above-mentioned scanning method. The methodology is called snapshot imaging spectroscopy. It has been used in image mapping spectroscopy tomography imaging spectroscopy and snapshot spectral imaging.

In techniques where a tuneable light source is used for the wavelength slicing, the detector is usually a simple industrial camera which can be a monochrome or colour camera. In reference [15] the authors have utilised a standard industrial camera DCM23UP1300(the Imaging source, Bremen Germany) for recording the interferograms as wavelength slicing was done at the source end. The various methods used for the acquisition involve the wavelength-slicing tool. In hyper-spectral imaging spectral tuneable filters, liquid crystal tuneable filter, and AOTF are mostly used.

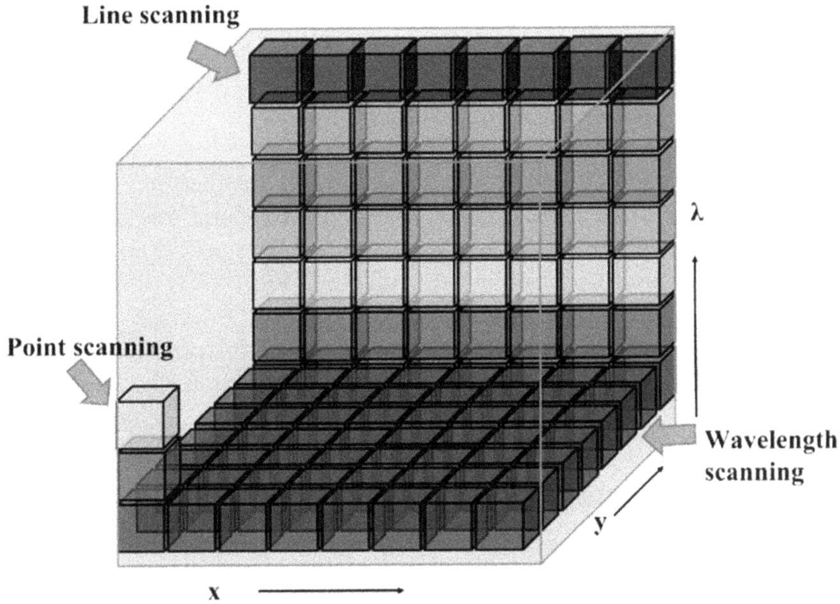

Figure 9.15. 3D hyper-spectral data cube with the data acquisition techniques point-scanning, line-scanning and wavelength scanning [2].

In reference [21], the authors have carried out AOTF-based hyper-spectral phase microscopy using a spatial light modulator (SLM) in combination with an AOTF. Their system was custom made using a wide angle paratellurite AOTF which had a tuning range of 600 to 1100 nm and a resolution of 20 cm^{-1}. They used a commercially available SLM having a resolution of 1920 × 1080 pixels with 8.5 micrometer pitch. The SLM was controlled using an HDMI interface and the images were captured using a monochrome camera (16 bit, with 6.8 micrometer pixel pitch).

The commercially available hyper-spectral recording devices include continuously varying tuneable filters and hyper-spectral cameras.

Continuously variable filters are developed in such a way that their spectral properties are varying along any one of the dimensions of the filter. These are wedged-shaped filter that help in adjusting the central wavelength of the filter just by sliding the filter in the varying dimensions. These filters are available in the range of linear variable longwave pass filter, linear variable shortwave pass filter and bandpass filter. They are also referred to as linear variable filter as they have an approximately linear relationship between the centre wavelength and distance along the filter.

In addition to this, there are various commercially available hyper-spectral cameras (HinaLea). These hyper-spectral cameras are based on the use of a front-steering Fabry–Pérot interferometer and do not require any mechanical scanning as in the case of the above-mentioned linear variable filters. They employ a tuneable filter placed in front of the sensor which can select the wavelength band sequentially and thus generate the complete hyper-spectral data cube. They offer high spectral and spatial resolution without any image uniformity problem compared to line-scanning methods.

Further, apart from hyper-spectral cameras, hyper-spectral microscope systems are also available which offer various advantages over the above technique. A multi-spectral microscopic camera uses a colour filter array on the focal plane or hyper-spectral microscope cameras which are grating-based require mechanical scanning. The HinaLea 4200M Microscope System is a steering hyper-spectral microscopy system which can scan in the visible–near infrared (VIS–NIR) range, i.e. 400 nm–1000 nm with high spatial and spectral resolutions. Another range of hyper-spectral cameras such as Specim FX have been designed for industrial and laboratory use which work on the methods of line scan. They are available in the VIS–NIR range, i.e. 400–100 nm.

9.7 Algorithms for image reconstruction

In multi-spectral imaging the image reconstruction can be carried out in multiple ways depending upon the imaging technique used during the acquisition of the images. If the fringe density is high enough then the image reconstruction can be carried out using Fourier transform method. If the acquisition is carried out using phase-shifting technique, then the reconstruction is to be carried out using phase-shifting algorithm itself.

9.7.1 Fourier transform method

The recorded intensity with the help of a CCD camera is given by the equation

$$I(x, y) = \mathrm{Iav}(x, y) + \mathrm{Imod}(x, y)\cos[\phi(x, y)] \tag{9.2}$$

where Iav(x,y) represents the average intensity, Imod (x, y) represents the modulated intensity, spatial coordinates are given by x and y, the phase change is signified by (x, y). The equation can be rewritten using Euler's formula as

$$I(x, y) = \mathrm{Iav}(x, y) + \mathrm{Imod}(x, y)/2\,[\exp(-j\varphi(x, y)) + \exp(+j\varphi(x, y))] \tag{9.3}$$

The bandpass filter is utilised to filter only 1 component such that the obtained filtered fringe approximation becomes

$$F(x, y) = \frac{\mathrm{Imod}(x, y)\exp(+j\varphi(x, y))}{2} \tag{9.4}$$

The phase of the sample is calculated by

$$\varphi(x, y) = \tan^{-1}\left(\frac{F\mathrm{img}(x, y)}{F\mathrm{real}(x, y)}\right) \tag{9.5}$$

The phase thus acquired is wrapped phase because of the discontinuities of 2π. The phase is unwrapped using any phase unwrapping method.

9.7.2 Phase-shifting method

(a) Five-step phase-shifting algorithm

The intensity of the recorded interferograms can be given by the formulae:

$$I_1(x, y, \lambda_i) = I_0(x, y, \lambda_i)\{1 + V_i\gamma(z)\cos[\varphi(x, y, \lambda_i) - 2\delta_i]\} \tag{9.6}$$

$$I_2(x, y, \lambda_i) = I_0(x, y, \lambda_i)\{1 + V_i\gamma(z)\cos[\varphi(x, y, \lambda_i) - \delta_i]\} \tag{9.7}$$

$$I_3(x, y, \lambda_i) = I_0(x, y, \lambda_i)\{1 + V_i\gamma(z)\cos[\varphi(x, y, \lambda_i)]\} \tag{9.8}$$

$$I_4(x, y, \lambda_i) = I_0(x, y, \lambda_i)\{1 + V_i\gamma(z)\cos[\varphi(x, y, \lambda_i) + \delta_i]\} \tag{9.9}$$

$$I_5(x, y, \lambda_i) = I_0(x, y, \lambda_i)\{1 + V_i\gamma(z)\cos[\varphi(x, y, \lambda_i) + 2\delta_i]\} \tag{9.10}$$

i stands for red, green, blue images. V is the visibility, and $\gamma(z)$ is the coherence envelop of the source. The phase can be obtained from the above intensity formula for $\delta = \frac{\pi}{2}$ by

$$\varphi(x, y, \lambda_i) = \tan^{-1}\left[\frac{2 *(I_2(x, y, \lambda_i) - I_4(x, y, \lambda_i))}{I_1(x, y, \lambda_i) - 2*I_3(x, y, \lambda_i) + I_5(x, y, \lambda_i)}\right] \tag{9.11}$$

In reference [5] the authors have employed the phase-shifting method in multi-spectral quantitative phase microscopy. They recorded five phase-shifted interferograms of human RBCs for each of the wavelengths used and reconstructed the phase profile of these RBCs for each of the wavelengths. The interferograms were recorded by the authors and the reconstructed phase profile is shown in figure 9.16. They used the obtained phase results to calculate the refractive index of the RBCs at different wavelength using the equation

$$\Delta\varphi(x, y, \lambda_i) = \frac{2\pi}{\lambda_i}[n_{\text{cell}}(x, y, \lambda_i) - n_{\text{med}}(x, y, \lambda_i)].\ 2h(x, y) \tag{9.12}$$

where n_{cell} and n_{med} are the refractive indexes of cell and medium. $h(x,y)$ is the thickness of the cell and the extra 2 term is added as the system operated in refection mode [5]. Figure 9.17 shows the phase and refractive index of RBC calculated.

(b) Spatial phase-shifting

In multi-spectral digital holography when an interferogram is recorded for each of the wavelengths, the reconstruction of the phase images can be carried out using a spatial phase-shifting algorithm as in references [22, 23]. It implies using non diffractive reconstruction method. The recorded hologram in the hologram plane is due to the superposition of reference and object waves and is given by the equation [23]

$$I_h(x, y,\ z = z_h) = I_0(x, y,\ z = z_h) + I_r(x, y,\ z = z_h) + 2\sqrt{I_o(x, y,\ z = z_h)I_r(x, y,\ z = z_h)}\cos\Delta\varphi_h(x, y,\ z = z_h) \tag{9.13}$$

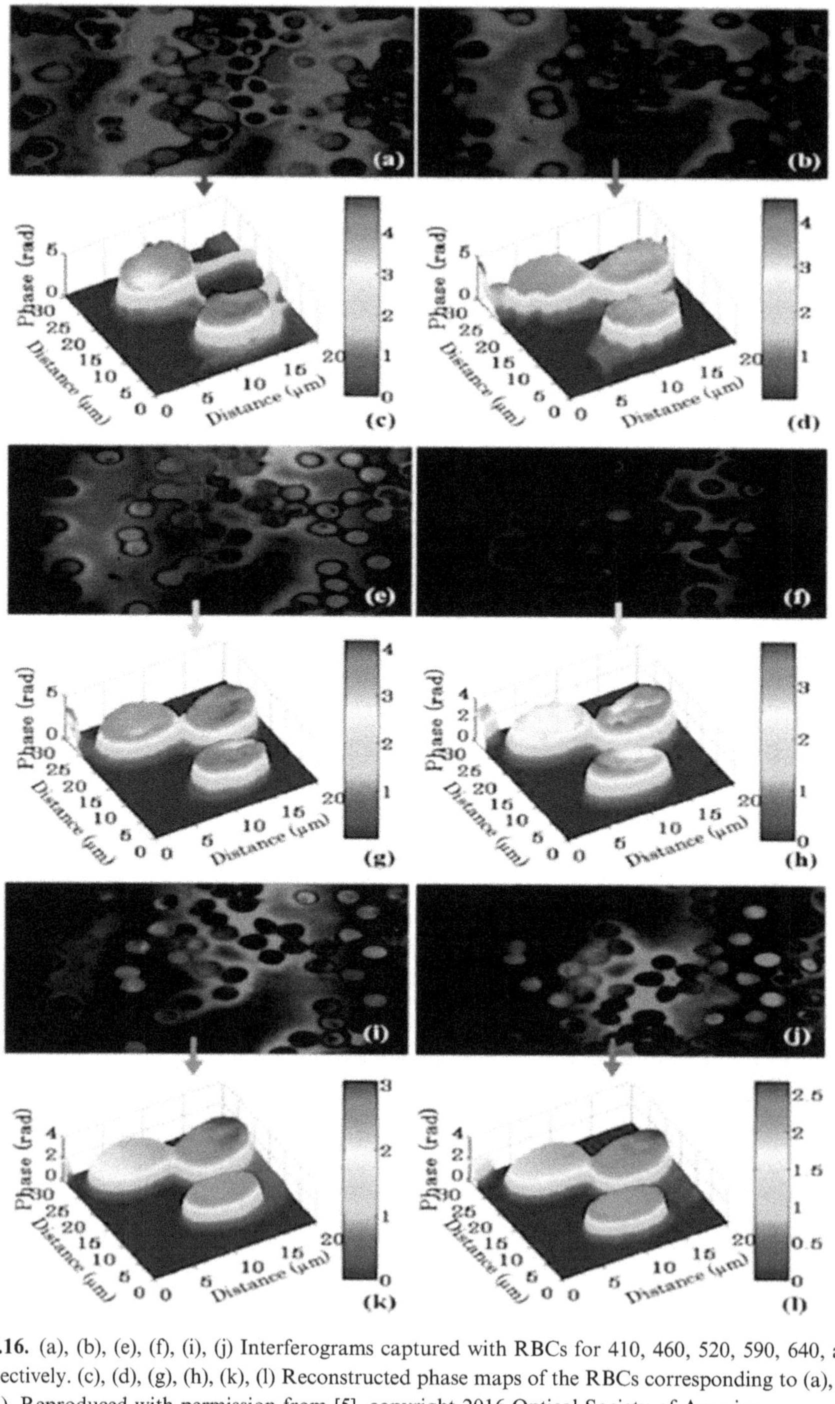

Figure 9.16. (a), (b), (e), (f), (i), (j) Interferograms captured with RBCs for 410, 460, 520, 590, 640, and 720 nm, respectively. (c), (d), (g), (h), (k), (l) Reconstructed phase maps of the RBCs corresponding to (a), (b), (e), (f), (i), (j). Reproduced with permission from [5], copyright 2016 Optical Society of America.

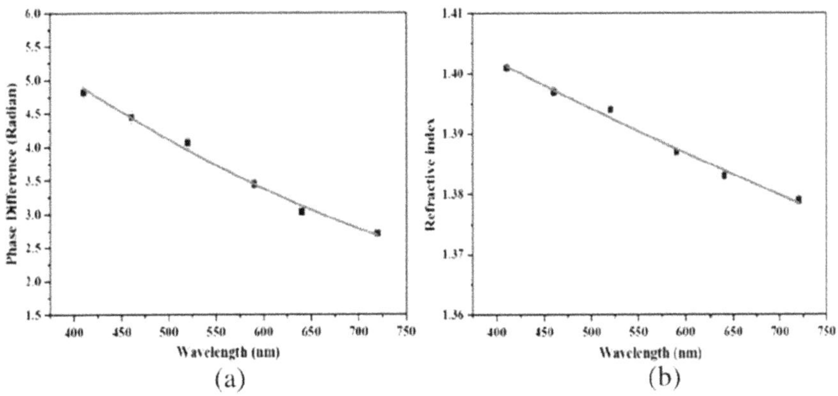

Figure 9.17. (a) Wavelength against phase shift of RBCs. (b) Refractive index variation of RBCs. Reproduced with permission from [5], copyright 2016 Optical Society of America.

In the above equation,

$$\Delta\varphi_h(x,\ y,\ z = z_h) = \ \varphi_r(x,\ y,\ z = z_h) - \ \varphi_o(x,\ y,\ z = z_h) \tag{9.14}$$

$$\Delta\varphi_h(x,\ y,\ z = z_h) = 2\pi\left(D_x x^2 + D_y y^2 + P_x x + P_y y\right) \tag{9.15}$$

In the above equation, the variables D_x and D_y are the divergence of the object wave, and P_x and P_y are the phase difference between object and reference waves. The quantitative phase is measured using a pixel-wise solution of equations (9.12) and (9.14).

For obtaining the solution, a square area of 5 × 5 neighbouring pixels around the given pixel is deemed by the usage of spatial space shifting algorithm [24, 25]. The proposed algorithm assumes that $\Delta\varphi_h$ generated at the hologram plane between object and reference wave varies rapidly and object wave intensity is constant in 5 × 5 neighbouring pixels of the given pixel. The variables in equation (9.14) cannot be obtained directly from the experimental set-up so they can be calculated using an iterative fitting process without the sample [24].

Thereafter, in order to obtain sharp images, further propagation of the object wave to the image plane z_i is carried out using Fresnel transformation.

$$\text{Obj}(x,\ y,\ z_i = z_h + \Delta z) = \ FT^{-1}\left\{FT[\text{Obj}(x,\ y,\ z_h)]e^{[i\pi\lambda\Delta z(v^2+u^2)]}\right\} \tag{9.16}$$

where wavelength of light used is given by λ in the experiment, FT is the Fourier transform and Δz is the distance at which the amplitude image is sharply focused.

In the case when no propagation is required the sharp image is obtained at $z_i = z_h$. In addition to amplitude, the phase information can be obtained from Obj$(x,\ y,\ z_i)$ using

$$\Delta\varphi(x,\ y,\ z_h) = \arctan\frac{\text{Img}\{\text{Obj}(x,\ y,\ z_i)\}}{\text{Real}\{\text{Obj}(x,\ y,\ z_i)\}} \tag{9.17}$$

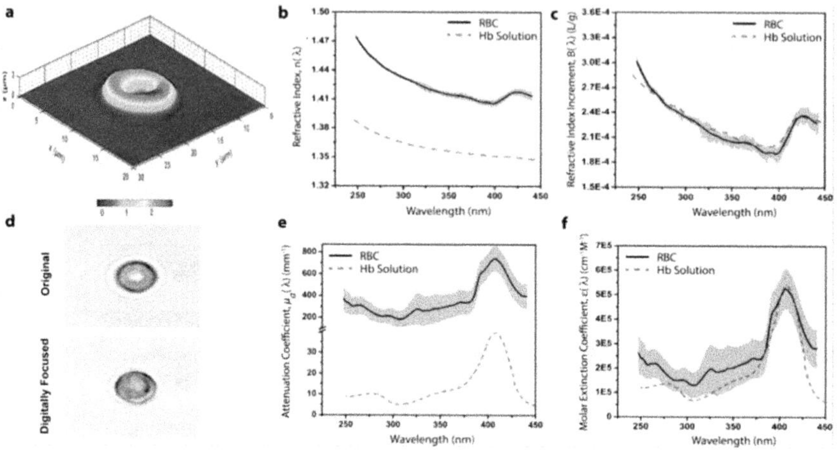

Figure 9.18 Ultraviolet hyper-spectral interferometric microscopy of living RBCs. (a) Topological estimated thickness map for RBC. (b) Refractive index for and 2 g dL^{-1} Hb solution given by black and dashed red line, respectively. (c) Real refractive index increment (B) spectra for the imaged RBC and Hb solution. (d) Digitally refocused amplitude map at 390 nm. (e) and (f) attenuation coefficient spectra and molar extinction coefficient spectra for the RBC and Hb solution. Reproduced with permission from [5], copyright 2018, The Author(s). With permission of Springer.

To the obtained phase of the sample, phase unwrapping is applied to remove the 2π ambiguity.

In another method which involves using an imaging spectrometer to acquire data that contains one spatial dimension as a function of wavelenth, the post processing starts by interpolating the data to wavenumber ($k = \frac{2\pi}{\lambda}$) and then applying 2D fast Fourier transform (FFT). The signal is filtered and shifted to the centre and inverse FFT is taken to give phase and amplitude in one direction. The above process is repeated at each 2D dataset to get a 3D complex hyper-spectral data cube [26]. The obtained complex field is propagated using Fresnel propagation to get focused amplitude and phase images at all the wavelengths. The complex signal obtained at each pixel location can be illustrated by the equation

$$I(\lambda) = I_0(\lambda).\ e^{-[u_{ab}(\lambda)+u_{sc}(\lambda)]L}.\ e^{-ikL[n(\lambda)]} \tag{9.18}$$

where $I_0(\lambda)$ is the initial intensity, u_{ab} and u_{sc} describe absorption and scattering terms, respectively, and the last term depends on the sample refractive index. The above equation describes that the obtained signal is directly related to absorption and scattering property of the specimen.

9.8 Applications of multi-spectral and hyper-spectral phase microscopy

Multi-spectral and hyper-spectral phase microscopy finds its application in the fields of biomedicine and biophotonics. Multi-spectral phase microscopy has been used in calculating the refractive index of biological specimens at various wavelengths. In reference [5] the authors used multi-spectral QPI to calculate the refractive index of human RBC at the wavelengths to 410, 460, 520, 590, 640, and 720 nm. It has also been used to calculate the refractive index distribution of mouse retina sections [22] as it a vital parameter which is used in ophthalmic imaging.

Hyper-spectral imaging was used in the fields of geoscience and remote sensing [28], phase microscopy, biomedicine and medical applications [29, 30]. Based on the acquisition strategy, various applications of hyper-spectral imaging have been developed such as point-scanning spectroscopy that is used in hyper-spectral confocal microscopy, push broom spectroscopy that is used in hyper-spectral line-scanning microscopy, while the third method of wavelength scanning is used in digital light processing-based imaging spectroscopy, liquid crystal tuneable filter-based imaging spectroscopy and acousto-optic based imaging spectroscopy [2].

Hyper-spectral confocal microscopy is a point-scanning method in which a hyper-spectral confocal microscope is used first to disperse the light in the spectral domain and then the images are captured with the help of a linear detector array. The hyper-spectral confocal microscope is very widely used in applications like live cell multiplexed imaging [31, 32]. It offers the advantage of intrinsic depth resolved spectral imaging in addition to reduced crosstalk in adjacent depth layers.

The hyper-spectral line-scanning microscope is based on the method of utilizing 2D detector arrays to collect the (Y, λ) slice of the data cube such that only one spatial dimension is required for scanning. It utilizes a line focusing lens in order to focus the excitation light into a uniform line. The fluorophores excited by the line excitation are collected by the same objective and acquisition takes place at the entrance of the slit of a spectrometer. The recorded line is dispersed into spectral components and taken using a 2D detector and scanning is accomplished along the length of the X axis to produce a three-dimensional data cube.

Application of a wavelength-scanning spectrometer involves the use of a tuneable filter-based imaging spectrometer. The main tuneable filters in this category include the use of liquid crystal tuneable filters [33, 34], AOTFs, filter arrays [35] and linear variable filters [36]. These filters are used in two emerging modalities: the first is referred to as excitation site tuneable filter; and the other is referred to as emission side tuneable filter. The data acquisition using tuneable filter-based imaging spectroscopy gives the advantage over point-scanning method as it is faster and requires scanning only in one dimension, that is, wavelength scanning. But it faces the disadvantage of lack of optical sectioning as they are unable to perform depth resolved imaging.

A snapshot imaging spectrometer has the advantage of capturing the entire 3D data cube in a single shot. It does not involve any scanning mechanism thus offering the advantage over other point-scanning or line-scanning methods. Many techniques have been developed based on snapshot imaging spectrometry, but they face the disadvantage of low-resolution imaging [37].

In conclusion, multi-spectral and hyper-spectral phase imaging find their application in the laboratory as well as in clinical studies. Their extension to the field of biomedicine helps us in gaining access to various parameters of biological samples. It helps in realizing new areas in the field of imaging and spectroscopy. Optical properties (absorption and scattering) of biological specimens are largely dependent on the wavelength of light and to understand their dependency on wavelength the data should be studied in the spectral domain.

References

[1] Jacques S L 2013 Optical properties of biological tissues: a review *Phys. Med. Biol.* **58** R37–61

[2] Gao L and Smith R T 2015 Optical hyperspectral imaging in microscopy and spectroscopy – a review of data acquisition *J. Biophoton.* **8** 441–56

[3] Pham H, Bhaduri B, Ding H and Popescu G 2012 Spectroscopic diffraction phase microscopy *Opt. Lett.* **37** 3438–40

[4] Kalenkov S G, Kalenkov G S and Shtanko A E 2013 Spectrally-spatial Fourier-holography *Opt. Express* **21** 24985–90

[5] Dubey V, Singh G, Singh V, Ahmad A and Mehta D S 2016 Multispectral quantitative phase imaging of human red blood cells using inexpensive narrowband multicolor LEDs *Appl. Opt.* **55** 2521–5

[6] Burmak L, Zykova L, Machikhin A and Chichigin B 2019 Multi-spectral phase imaging based on acousto-optic selection of light in common-path interferometer *J. Phys. Conf. Ser.* **1421** 012055

[7] Mehta D S and Srivastava V 2012 Quantitative phase imaging of human red blood cells using phase-shifting white light interference microscopy with colour fringe analysis, *Appl. Phys. Lett.* **101** 203701

[8] Ahmad A, Kumar A, Dubey V, Butola A, Ahluwalia B S and Mehta D S 2019 Characterization of color cross-talk of CCD detectors and its influence in multispectral quantitative phase imaging *Opt. Express* **27** 4572–89

[9] Dubey V, Singh V, Ahmad A, Singh G and Mehta D S 2016 White light phase shifting interferometry and color fringe analysis for the detection of contaminants in water *Quantitative Phase Imaging II (Proc. SPIE)* 97181F–7181

[10] Butola A, Ahmad V, Dubey P, Senthilkumaran and Mehta D S 2018 Spectrally resolved laser interference microscopy, *Laser Phys. Lett.* **15** 075602

[11] Mehta D S, Sharma A, Dubey V, Singh V and Ahmad A 2016 Quantitative phase imaging of biological cells and tissues using singleshot white light interference microscopy and phase subtraction method for extended range of measurement *Quantitative Phase Imaging II* vol 9718 (Bellingham, WA: SPIE) pp 257–64

[12] Dubey V, Ahmad A, Singh V, Butola A and Mehta D S 2017 Quantitative phase imaging using white light interference microscopy with color fringe analysis; a comparative study of color interferograms recorded by single chip and 3-chip CCD color camera *European Conference on Biomedical Optics* (Washington, DC: Optical Society of America) 104140I

[13] Mann P, singh V, Tayal S and Mehta D S 2020 White light interference microscopy with color fringe analysis for quantitative phase imaging and 3-D step height measurement *Imaging and Applied Optics Congress, OSA Technical Digest* JW2A.13 Optica Publishing Group

[14] Mann P, Singh V, Tayal S, Thapa P and Mehta D S 2022 White light phase shifting interferometric microscopy with whole slide imaging for quantitative analysis of biological samples *J. Biophoton.* **15** e202100386

[15] Barroso Á *et al* 2018 Hyperspectral digital holographic microscopy approach for reduction of coherence induced disturbances in quantitative phase imaging of biological specimens *Proc. SPIE* **10834** 108340I-1

[16] Chen S, Ryu J, Lee K and Zhu Y 2016 Swept source digital holographic phase microscopy *Opt. Lett.* **41** 665–8
[17] Butola A *et al* 2017 Quantitative phase imaging using spectrally resolved white light interferometry *Proc. SPIE* **104140G**
[18] Ba C, Tsang J-M and Mertz J 2018 Fast hyperspectral phase and amplitude imaging in scattering tissue *Opt. Lett.* **43** 2058
[19] Yushkov K B and Molchanov V Y 2017 Hyperspectral imaging acousto-optic system with spatial filtering for optical phase visualization *J. Biomed. Opt.* **22** 1–7
[20] https://olympuslifescience.com/en/microscoperesource/primer/techniques/confocal/aotfintro/
[21] Yushkov K B, Champagne J, Kastelik J-C, Makarov O Y and Molchanov V Y 2020 AOTF-based hyperspectral imaging phase microscopy *Biomed. Opt. Express* **11** 7053
[22] Peña Á B *et al* 2019 Refractive index properties of the retina accessed by multi-wavelength digital holographic microscopy *Proc. SPIE* **10883**
[23] Kemper B *et al* 2006 Investigation of living pancreas tumor cells by digital holographic microscopy *J. Biomed. Opt.* **11** 1–8
[24] Carl D, Kemper B, Wernicke G and von Bally G 2004 Parameter-optimized digital holographic microscope for high-resolution living-cell analysis *Appl. Opt.* **43** 6536
[25] Liebling M, Blu T and Unser M 2004 Complex-wave retrieval from a single off-axis hologram *J. Opt. Soc. Am. A* **21** 367
[26] Robles F E 2021 Deep-UV hyperspectral microscopy with phase and absorption-based contrast *Proc. SPIE* **11653**
[27] Ojaghi A, Fay M E, Lam W A and Robles F E 2018 Ultraviolet hyperspectral interferometric microscopy *Sci. Rep.* **8** 9913
[28] Lee Z, Carder K L, Mobley C D, Steward R G and Patch J S 1998 Hyperspectral remote sensing for shallow waters I a semianalytical model *Appl. Opt.* **37** 6329
[29] Godden T M, Muñiz-Piniella A, Claverley J D, Yacoot A and Humphry M J 2016 Phase calibration target for quantitative phase imaging with ptychography *Opt. Express* **24** 7679
[30] Katkovnik V, Shevkunov I, Petrov N v and Egiazarian, K 2016 High-accuracy off-axis wavefront reconstruction from noisy data: local least square with multiple adaptive windows *Opt. Express* **24** 25068
[31] J V W F *et al* 2008 *In vivo* hyperspectral confocal fluorescence imaging to determine pigment localization and distribution in cyanobacterial cells *Proc. Natl Acad. Sci.* **105** 4050–5
[32] Haaland D M *et al* 2007 Hyperspectral confocal fluorescence imaging of cells *Proc. SPIE* **6765**
[33] Lansford R, Bearman G H and Fraser S E 2001 Resolution of multiple GFP color variants and dyes using two-photon microscopy and imaging spectroscopy *J. Biomed. Opt.* **6** 311–8
[34] Akbari H *et al* 2012 Hyperspectral imaging and quantitative analysis for prostate cancer detection *J. Biomed. Opt.* **17** 1–11
[35] Favreau P F *et al* 2014 Excitation-scanning hyperspectral imaging microscope *J. Biomed. Opt.* **19** 1–11
[36] di Caprio G, Schaak D and Schonbrun E 2013 Hyperspectral fluorescence microfluidic (HFM) microscopy *Biomed. Opt. Express* **4** 1486
[37] Hagen N A and Kudenov M W 2013 Review of snapshot spectral imaging technologies *Opt. Eng.* **52** 1–23

IOP Publishing

Quantitative Phase Microscopy and Tomography
Techniques using partially spatially coherent monochromatic light
Dalip Singh Mehta, Ankit Butola and Veena Singh

Chapter 10

Conclusions and future prospects

Quantitative phase microscopy (QPM) [1–8] has become indispensable technology for label-free quantitative analysis of various biological cells and tissues, such as sperm cells [5, 6, 9], macrophages, liver sinusoidal cells [8], red blood cells [1–8], and cancerous cells [10, 11], without using any exogenous contrast agents. QPM generates high contrast imaging of nearly transparent specimens using their intrinsic refractive index distribution and hence optical path length (OPL) change, leading to change in the phase values resulting in significant change in intensity variation across the sample space [1–12]. This significant change in OPL and hence phase of light can be converted into significant contrast change via various QPM configurations, and hence both high contrast imaging and quantitative information about the sample can be extracted [1–5]. The key parameters controlling measurement accuracy and capability of any QPM system depends on its spatial and temporal phase sensitivity. The spatial phase sensitivity of QPM is governed by the coherence properties of the light source [12–17] and temporal stability depends on the optical configuration of QPM whether it is common-path or off-axis in nature [17–19]. Quantification of various morphological and biophysical parameters can be utilized for early-stage disease detection, reproductive science (*in-vitro* fertilization) and infectious diseases [1–19]. Precise determination of these parameters depends on the accuracy in phase measurement of the system, which is highly affected by the spatial phase noise and temporal phase stability of the interferometric system utilized for QPM. An ideal QPM system incorporates low spatial phase noise and high temporal phase stability leading to high accuracy in phase measurement. This depends on the two important parts of the QPM system: first the use of a suitable light source without any speckle noise, spurious fringes; and secondly the optical interferometric configuration which guarantees the stability of the hologram or interferogram. Most QPM techniques utilize highly coherent light sources like lasers, benefiting from their remarkable properties, such as high spatial and temporal coherence,

doi:10.1088/978-0-7503-3987-2ch10

directionality, monochromaticity and brightness [1–21]. These remarkable properties of laser light guarantee the setting-up of an interferometer system quickly and recording high visibility fringes. High monochromaticity of lasers also guarantees no dispersion effects. On the other hand, laser's high spatio-temporal coherence leads to occurrence of speckle noise and spurious fringes leading to inhomogeneous illumination and poor spatial phase sensitivity [12–21]. All these factors significantly influence the illumination, image quality and therefore phase measurement accuracy of QPM systems. In this book we have mainly demonstrated and described the QPM techniques with partially spatially coherent monochromatic (PSCM) and low coherent light. During the last few years, various QPM systems have been developed using PSCM light sources and demonstrated an improvement in the phase measurement accuracy [12–31]. Interestingly, we see an order of magnitude reduced spatial phase noise compared to coherent lasers [17]. Finally, a high space-bandwidth product with excellent spatial phase sensitivity and temporal phase sensitivity is important for QPM technology. In the last year we have also demonstrated a highly stable speckle-free QPM system in which we could achieve super stable fringe pattern leading to an order of magnitude higher temporally stable compared to conventional off-axis systems. This system can be utilized for precise measurement of dynamic membrane fluctuation studies.

We have also demonstrated quantitative phase imaging with high spatial phase sensitivity and implemented a deep neural network for the classification of various clinically relevant applications such as human spermatozoa cells under stressed condition [6, 9]. Quantitative measurement of various parameters, such as, morphology and sub-cellular changes in macrophages and trophoblasts during inflammation [28] has also been conducted. A PSCM based QPM system is able to provide sperm cell motility, morphology, and quantitative analysis with high accuracy and we could distinguish normal sperm cells from cells under different stress conditions such as cryopreservation, exposure to hydrogen peroxide and ethanol, which was not possible with coherent light based QPM [28]. The reduced spatial phase noise allowed us to reconstruct precisely the complete tail, neck and head part of a full sperm cell, which was missing in coherent light source (see chapter 2). Further, we have also investigated the effect of pro-inflammatory cytokines such as TNF-α and INF-γ on the macrophages and trophoblasts using QPM technique. In pregnancy during an inflammatory condition, macrophages present at the fetomaternal junction release an increased amount of nitric oxide (NO), which can disturb the trophoblast functions and pregnancy outcome. Precise determination of morphological modifications of cellular and sub-cellular structures associated with inflammatory responses are important to further quantify the extent of trophoblast dysfunction for clinical implications.

In chapters 1 and 2 we have demonstrated that the coherence properties of the light sources play an important role in QPM as they influence the measurement capability and accuracy of the system. Particularly, the effect of spatial coherence properties of laser light source on the performance of QPM technique is investigated in detail. These properties are modified by passing a coherent laser beam through an optically rough screen (diffuser), which generates static speckle patterns at its output. The diffuser is then either vibrated or rotated to suppress the speckle pattern effectively. The randomly scattered light is collected by a lens and focused onto a controllable aperture, so that the partial spatial coherence of the laser light can be controlled. Two schemes for generating spatially partially coherent light are implemented: (1) coherent laser beam + rotating diffuser; and (2) coherent laser beam + stationary diffuser + vibrating multiple multi-mode fiber bundle (MMFB). The PSCM light is then made incident onto the interferometers; off-axis and common-path configurations. In chapter 2 we have demonstrated the off-axis QPM system using partially spatially coherent light. Significant improvement in phase measurement has been achieved by means of using PSCM light compared to coherent laser. Another important aspect to consider in QPM is the temporal phase stability of the system along with the spatial phase sensitivity. Off-axis QPM systems (chapter 2) are very sensitive to external vibrations and therefore the phase measurement is affected by the temporal instability of the system. To solve this problem chapter 3 is dedicated to common-path QPM systems using PSCM [18, 19, 32–34]. We observed that temporal phase sensitivity of the common-path QPM system is an order of magnitude better than the off-axis QPM systems. We have demonstrated that, by means of taking into account both the spatial phase sensitivity and temporal phase sensitivity that are important for a QPM system, we conducted various experiments using common-path QPM system with PSCM light, thus solving the main problems of QPM systems. In summary, we have found that the use of PSCM light and a common-path QPM system provides speckle-free imaging, high spatial phase sensitivity and high temporal phase stability, leading to high accuracy in the phase/height measurement of the specimens. Therefore, a QPM system with common-path optical configuration while using PSCM light can be the most suitable system for precise measurement of quantitative information about biophysical parameters.

Almost all the QPM systems (off-axis or common-path configurations) also suffer from diffraction limited spatial resolution, therefore, the smallest object size that can be visualized using these systems always $\geqslant \lambda/2$. In order to improve the spatial resolution of QPM systems, there have been many attempts by researchers to break the diffraction limited resolution of the QPM system using structured illumination. In chapter 4, we describe structured illumination quantitative phase microscopy (SIQPM) [35–41]. Both off-axis and common-path SIQPM systems are described. By means of using SIQPM a twofold improvement in spatial resolution is observed. Multi-modal chip- based nanoscopy and quantitative phase microscopy have also been demonstrated and details are described in chapter 5. In this system both fluorescence and QPM have been integrated. The waveguide structure for total internal feftection (TIR) was used for fluorescence microscopy also known as TIRF and Linnik-type QPM system was used for phase microscopy. By means of TIRF one can obtain super-resolution and large field of view simultaneously, in addition to speckle-free QPM using PSCM. Using this method we were able to reconstruct the super-resolution image of liver sinusoidal endothelial cells (LSECs). In LSECs the sizes of the structures and fenestration are in the range between 50 and 150 nm [8]. The details are discussed in chapter 5.

Temporally low coherence (broad-band) light plays an important role in optical metrology (white light profilometers), optical coherence tomography (OCT), ranging, eye length measurement and sensing. In chapters 6 and 7 the use of low coherence light for OCT, and phase-shifting white light interferometry (PS-WLI) with color fringe analysis for QPM have been described [42–46]. We have also demonstrated that the PSCM light can be used for optical sectioning of multi-layer samples by means of using the longitudinal spatial coherence of the light source. The experimental results are demonstrated in chapter 6. Therefore, by means of using an extended PSCM light source one can achieve speckle-free imaging (free from coherent noise) without dispersion compensation, with high phase sensitivity and high-resolution optical sectioning as well. Further, by means of using WLI with color fringe analysis one can perform QPM with multiple wavelengths simultaneously. This leads to the determination of refractive index of the sample as a function of wavelength of light, however, the samples must be made in two refractive index media or a sample with known height. In chapter 7 we have demonstrated PS-WLI with color fringe analysis for red blood cells, cancer cells and *Escherichia coli* bacteria. Further, by means of using white light with a common-path interferometer (Mirau-type objective lens) one can obtain both high spatial and high temporal phase sensitivity along with speckle-free high contrast imaging. [2, 52–57]. In table 10.1, we summarize all the merits and demerits of various QPM systems using incoherent, coherent and partially spatially coherent monochromatic light. We find that QPM with PSCM light source is the most suitable system with maximum merits.

Finally, the road map for the future of QPM is the use of various computational techniques such as machine learning, support vector machine (SVM), convolutional neural network (CNN) and artificial intelligence (AI). By means of using these advanced computational tools along with the aforementioned QPM techniques, early diagnosis of diseases can be done with high accuracy and high sensitivity. In chapter 8 we describe AI: a computational tool to interpret quantitative phase imaging data. Further, deep learning (DL)-based phase recovery, virtual staining and resolution enhancement have a broad application in biomedical imaging, data interpretation, material science, among others. Therefore, rigorous research is ongoing for QPM with AI, ML and DL combined, which has a great future.

In addition, the QPM techniques are being implemented using multi-spectral and hyper-spectral concepts. Some details are given in chapter 9 on multi-spectral and hyper-spectral quantitative phase microscopy. By means of performing QPM with multi-spectral and hyper-spectral techniques along with implementing AI, ML and DL, rich information about the biological samples can be gathered. Further, multi-spectral and hyper-spectral QPM is always demonstrated using broad-band light and hence is free from speckle and spurious fringes. Therefore, QPM techniques have a great future.

Table 10.1. Comparison of different types QPM systems, light sources and performance parameters. Green color: merits; red color: demerits.

Green color: Merits
Red color: Demerits

	Low coherence light (Halogen lamp / LED)	Coherent Light (Lasers)	PSCM light
Interferometer Types that can be used for QPM	Preferably common path	Both off axis as well as common path can be used	Both off axis as well as common path can be used
Temporal coherence length	Extremely short	Too long	Long but longitudinal spatial coherence can play role
Free from spurious fringes and speckle noise	Yes	No	Yes
Spatial phase sensitivity	High	Low	High
Temporal phase sensitivity	High	Low for off axis and high for common path	Low for off axis and high for common path
High contrast fringes over full field of view (FOV) of camera	Only for small FOV	Over large FOV	Over large FOV
Space bandwidth product	Low	High	High
Measurement accuracy (phase, OPL and height)	High	Low	High
Chromatic aberration	Yes	No	No
Dispersion compensation	Required	Not required	Not required
Dynamic study of biological cells	Possible if operated in single shot or with parallel phase shifting	Possible	Possible
Tomography (Optical Sectioning)	Possible with dispersion compensation	Not possible	Possible using LSC and without dispersion compensation
Degeneracy (No of photons per unit area per unit time)	Low	High	Moderate

References

[1] Popescu G 2011 *Quantitative Phase Imaging of Cells and Tissues* (New York: McGraw-Hill Education)

[2] Mehta D S 2018 Quantitative phase optical microscopic techniques for biomedical imaging and diagnostic applications *Proc. Natl. Acad. Sci., India, Sect. A Phys. Sci.* **88** 437–51

[3] Choi W, Fang-Yen C, Badizadegan K, Oh S, Lue N, Dasari R R and Feld M S 2007 Tomographic phase microscopy *Nat. Methods* **4** 717–19

[4] Mehta D S and Srivastava V 2012 Quantitative phase imaging of human red blood cells using phase-shifting white light interference microscopy with colour fringe analysis *Appl. Phys. Lett.* **101** 203701–3
[5] Butola A *et al* 2020 High spatially sensitive quantitative phase imaging assisted with deep neural network for classification of human spermatozoa under stressed condition *Sci. Rep.* **10** 1–12
[6] Atzitz Y, Dudaie M, Barnea I and Shaked N T 2021 Sperm inspection for *in vitro* fertilization via self-assembled microdroplet formation and quantitative phase microscopy *Cells* **10** 3317
[7] Dubey V, Singh G, Singh V, Ahmad A and Mehta D S 2016 Multi-spectral quantitative phase imaging of human red blood cells using inexpensive narrow band multi-color LEDs *Appl. Opt.* **55** 2521–5
[8] Butola A *et al* 2022 Multimodal on-chip nanoscopy and quantitative phase imaging reveals the nanoscale morphology of liver sinusoidal endothelial cells *Proc. Natl Acad. Sci.* **118** e2115323118
[9] Dubey V, Popova D, Ahmad A, Acharya G, Basnet P, Mehta D S and Ahluwalia B S 2019 Partially spatially coherent digital holographic microscopy and machine learning for quantitative analysis of human spermatozoa under oxidative stress condition *Sci. Rep.* **9** 3564
[10] Baruch S B, Rotman-Nativ N, Baram A, Greenspan H and Shaked N T 2021 Cancer-cell deep-learning classification by integrating quantitative-phase spatial and temporal fluctuations *Cells* **10** 3353
[11] Dubey V, Ahmad A, Butola A, Qaiser D, Srivastava A and Mehta D S 2019 Low coherence quantitative phase microscopy with machine learning model and Raman spectroscopy for the study of breast cancer cells and their classification, *Appl. Opt.* **58** A112–9
[12] Dubois F, Joannes L and Legros J-C 1999 Improved three-dimensional imaging with a digital holography microscope with a source of partial spatial coherence *Appl. Opt.* **38** 7085–94
[13] Dubois F, Requena M-L N, Minetti C, Monnom O and Istasse E 2004 Partial spatial coherence effects in digital holographic microscopy with a laser source *Appl. Opt.* **43** 1131–9
[14] Dubois F, Callens N, Yourassowsky C, Hoyos M, Kurowski P and Monnom O 2006 Digital holographic microscopy with reduced spatial coherence for three-dimensional particle flow analysis, *Appl. Opt.* **45** 864–71
[15] Kolman P and Chmelík R 2010 Coherence-controlled holographic microscope *Opt. Express* **18** 21990–2003
[16] Pan F, Xiao W, Liu S, Wang F J, Rong L and Li R 2011 Coherent noise reduction in digital holographic phase contrast microscopy by slightly shifting object *Opt. Express* **19** 3862–9
[17] Ahmad A, Dubey V, Jayakumar N, Habib A, Butola A, Nystad M, Acharya G, Basnet P, Mehta D S and Ahluwalia B S 2021 High-throughput spatial sensitive quantitative phase microscopy using low spatial and high temporal coherent illumination *Sci. Rep.* **11** 1–13
[18] Singh V, Tayal S and Mehta D S 2019 Speckle free common path digital interference phase microscopy using single element interferometers with partially spatially coherent light source *Asian J. Phys.* **28** 10–2
[19] Tayal S, Singh V, Kaur T, Singh N and Mehta D S 2020 Simultaneous fluorescence and quantitative phase imaging of MG63 osteosarcoma cells to monitor morphological changes with time using partially spatially coherent light source *Methods. Appl. Fluoresc.* **8** 035004

[20] Bianco V, Memmolo P, Leo M, Montresor S, Distante C, Paturzo M, Picart P, Javidi B and Ferraro P 2018 Strategies for reducing speckle noise in digital holography *Light: Sci. Appl.* **7** 48

[21] Redding B, Choma M A and Cao H 2012 Speckle-free laser imaging using random laser illumination *Nat. Photonics* **6** 355–9

[22] Edwards C, Bhaduri B, Nguyen T, Griffin B G, Pham H, Kim T, Popescu G and Goddard L L 2014 Effects of spatial coherence in diffraction phase microscopy *Opt. Express* **22** 5133–46

[23] Rodrigo J A and Alieva T 2014 Rapid quantitative phase imaging for partially coherent light microscopy *Opt. Express* **22** 13472–83

[24] Rodrigo J A and Alieva T 2014 Illumination coherence engineering and quantitative phase imaging *Opt. Lett.* **39** 5634–7

[25] Nguyen T H, Edwards C, Goddard L L and Popescu G 2014 Quantitative phase imaging with partially coherent illumination *Opt. Lett.* **39** 5511–4

[26] Nguyen T H, Edwards C, Goddard L L and Popescu G 2016 Quantitative phase imaging of weakly scattering objects using partially coherent illumination *Opt. Express* **24** 11683–93

[27] Butola A, Kanade S R, Bhatt S, Dubey V K, Kumar A, Ahmad A, Prasad D K, Senthilkumaran P, Ahluwalia B S and Mehta D S 2020 High space-bandwidth in quantitative phase imaging using partially spatially coherent optical coherence microscopy and deep neural network *Opt. Express* **28** 36229–44

[28] Singh R, Dubey V, Wolfson D, Ahmad A, Butola A, Acharya G, Mehta D S, Basnet P and Ahluwalia B S 2020 Quantitative assessment of morphology and sub-cellular changes in macrophages and trophoblasts during inflammation *Biomed. Opt. Express* **11** 3733–52

[29] Ahmad A, Dubey A V, Butola A, Tinguely J C, Ahluwalia B S and Mehta D S 2020 Sub-nanometer height sensitivity by phase shifting interference microscopy under environmental fluctuations *Opt. Express* **28** 9340–58

[30] Nikhil Jayakumar B S A, Ahmad A, Mehta D S and Ahluwalia B S 2020 Sampling moiré method: a tool for sensing quadratic phase distortion and its correction for accurate quantitative phase microscopy *Opt. Express* **28** 10062–77

[31] Usmani K, Ahmad A, Joshi R, Dubey V, Butola A and Mehta D S 2019 Relationship between the source size at the diffuser plane and the longitudinal spatial coherence function of the optical coherence microscopy system *JOSA* A **36** D41–6

[32] Tayal S, Saxena A, Singh V, Kaur T, Singh N, Khare K and Mehta D S 2022 Design and development of integrated TIRF and common-path quantitative phase microscopic health care system with high stability *Opt. Lasers. Eng.* **155** 107057

[33] Tayal S, Singh V, Kaur T, Singh N and Mehta D S 2022 Multimodal biomicroscopic system for the characterization of cells with high spatial phase sensitivity and sub-pixel accuracy *J. Biophoton.* **15** e202100258

[34] Ahmad, Dubey V, Butola A, Ahluwalia B S and Mehta D S 2020 Highly temporal stable, wavelength-independent, and scalable field-of-view common-path quantitative phase microscope *J. Biomed. Opt.* **25** 116501

[35] Neumann Y, Kuznetsova and Brueck S R J 2008 Structured illumination for the extension of imaging interferometric microscopy *Opt. Express* **16** 6785–93

[36] Gao P, Pedrini G and Osten W 2013 Phase retrieval with resolution enhancement by using structured illumination *Opt. Lett.* **38** 5204–7

[37] Lai X J, Tu H Y, Lin Y C and Cheng C J 2018 Coded aperture structured illumination digital holographic microscopy for superresolution imaging *Opt. Lett.* **43** 1143–146

[38] Chowdhury S and Izatt J 2013 Structured illumination quantitative phase microscopy for enhanced resolution amplitude and phase imaging *Biomed. Opt. Express* **4** 1795–805
[39] Chowdhury S and Izatt J 2014 Structured illumination diffraction phase microscopy for broadband, sub-diffraction resolution, quantitative phase imaging *Opt. Lett.* **39** 1015–8
[40] Bhatt S, Butola A, Kanade S R, Kumar A and Mehta D S 2021 High-resolution single-shot phase-shifting interference microscopy using deep neural network for quantitative phase imaging of biological samples *J. Biophoton.* **14** e202000473
[41] Kumar V, Dubey A K, Gupta M, Singh V, Butola A and Mehta D S 2021 Speckle noise reduction strategies in laser-based projection imaging, fluorescence microscopy, and digital holography with uniform illumination, improved image sharpness, and resolution *Opt. Laser Technol.* **141** 107079
[42] Butola A, Prasad D K, Ahmad A, Dubey V, Qaiser D, Srivastava A, Senthilkumaran P, Ahluwalia B S and Mehta D S 2020 Deep learning architecture lightOCT for diagnostic decision support using optical coherence tomography images of biological samples *Biomed. Opt. Express* **11** 5017–31
[43] Singh V, Srivastava V and Mehta D S 2020 Machine learning-based screening of red blood cells using quantitative phase imaging with micro-spectrocolorimetry *Opt. Laser Technol.* **124** 105980
[44] Ahmad A, Kumar A, Dubey V, Butola A, Ahluwalia B S and Mehta D S Characterization of color cross-talk of CCD detectors and its influence in multispectral quantitative phase imaging *Opt. Express* **27** 4572–89
[45] Singh V, Joshi R, Tayal S and Mehta D S 2019 Speckle-free common-path quantitative phase imaging with high temporal phase stability using a partially spatially coherent multi-spectral light source *Laser Phys. Lett.* **16** 025601
[46] Butola A, Ahmad A, Dubey V, Srivastava V, Qaiser D, Srivastava A, Senthilkumaran P and Mehta D S 2019 Volumetric analysis of breast cancer tissue using machine learning and swept source optical coherence tomography *Appl. Optics* **58** A135–41
[47] Ahmad A, Mahanty T, Dubey V, Butola A, Ahluwalia B S and Mehta D S 2019 Effect on the longitudinal coherence properties of pseudo thermal light source as a function of source size and temporal coherence *Opt. Lett.* **44** 1817–20
[48] Ahmad A, Srivastava V, Dubey V and Mehta D S 2015 Ultra-short longitudinal spatial coherence length of laser light with the combined effect of spatial, angular, and temporal diversity *Appl. Phys. Lett.* **106** 093701
[49] Soto J M, Rodrigo J A and Alieva T 2017 Label-free quantitative 3D tomographic imaging for partially coherent light microscopy *Opt. Express* **25** 15699–712
[50] Native and Shaked N T 2017 Compact interferometric module for full-field interferometric phase microscopy with low spatial coherence illumination *Opt. Lett.* **42** 1492–4
[51] Kumar V, Dubey A K, Gupta M and Mehta D S 2022 Speckle free efficient light engine for high power laser projectors and automobile headlamps *Opt. Laser Technol.* **152** 108145
[52] Srivastava V, Inam M, Kumar R and Mehta D S 2016 Single shot white light interference microscopy for 3D surface profilometry using single chip color camera, *J. Opt. Soc. Korea* **20** 784–93
[53] Ahmad A, Dubey V K, Singh G, Singh V and Singh Mehta D 2016 Quantitative phase imaging of biological cells using spatially low and temporally high coherent light source *Opt. Lett.* **41** 1554–7

[54] Tayal S, Usmani K, Singh V, Dubey V and Mehta D S 2019 Speckle-free quantitative phase and amplitude imaging using common-path lateral shearing interference microscope with pseudo-thermal light source illumination *Optik* **177** 991–6
[55] Dubey V, Ahmad A, Singh R, Wolfson D L, Basnet P, Acharya G, Mehta D S and Ahluwalia B S 2018 Multi-modal chip-based fluorescence and quantitative phase microscopy for studying inflammation in macrophages *Opt. Express* **26** 19864–76
[56] Butola, Joshi T, Ahmad A, Dubey V, Senthilkumaran P and Mehta D S 2018 3D topography and tomography of multilayered freeform optical surfaces using swept source low coherence interferometry with large range of measurement *Laser Phys.* **28** 116101
[57] Butola, Ahmad A, Dubey V, Senthilkumaran P and Mehta D S 2018 Spectrally resolved laser interference microscopy *Laser Phys. Lett.* **15** 075602

www.ingramcontent.com/pod-product-compliance
Ingram Content Group UK Ltd.
Pitfield, Milton Keynes, MK11 3LW, UK
UKHW052041301026
9729IPUK00019B/346